Manufacturing Technology 3

P. J. Harris TEng(CEI), AMIProdE, MIQA

Lecturer in the Department of Engineering, Woolwich College

THE BUTTERWORTH GROUP

UNITED KINGDOM	Butterworth & Co (Publishers) Ltd London: 88 Kingsway, WC2B 6AB
AUSTRALIA	Butterworths Pty Ltd Sydney: 586 Pacific Highway, Chatswood, NSW 2067 Also at Melbourne, Brisbane, Adelaide and Perth
CANADA	Butterworth & Co (Canada) Ltd Toronto: 2265 Midland Avenue, Scarborough, Ontario M1P 4S1
NEW ZEALAND	Butterworths of New Zealand Ltd Wellington: 31–35 Cumberland Place CPO Box 472
SOUTH AFRICA	Butterworth & Co (South Africa) (Pty) Ltd Durban: 152–154 Gale Street
USA	Butterworth (Publishers) Inc Boston: 10 Tower Office Park, Woburn, Mass. 01801

First published 1981

© Butterworth & Co (Publishers) Ltd, 1981

All rights reserved. No part of this publication may be reproduced or transmitted in any form or by any means, including photocopying and recording, without the written permission of the copyright holder, application for which should be addressed to the Publishers. Such written permission must also be obtained before any part of this publication is stored in a retrieval system of any nature.

This book is sold subject to the Standard Conditions of Sale of Net Books and may not be re-sold in the UK below the net price given by the Publishers in their current price list.

British Library Cataloguing in Publication Data

Harris, P J
    Manufacturing technology 3.
    1. Manufacturing processes
    I. Title
    670   TS183

ISBN 0-408-00493-2

Typeset by Page Bros Ltd, Norwich
Printed and Bound by
Trade Litho Book Printers Ltd, Bodmin, Cornwall

Preface

This book, the second of two volumes, has been written to cover the work required for manufacturing technology at level 3 of the A5 Mechanical and Production Engineering programme of the Technician Education Council (TEC). The style and layout of this book are similar to *Manufacturing Technology 2* in the same series, emphasis being laid on principles rather than a detailed account of how a process should be carried out from a craft point of view. During the course of preparation I have included some topics that do not appear in the unit for this subject (U76/057). These additions include the dividing head and helical milling which I feel might well be incorporated if the unit is revised at some future date.

A subject such as manufacturing technology covers a very broad spectrum of engineering activity and it would be extremely difficult, if not impossible, to cover every aspect. However, it is hoped that this volume, together with the previous volume, will not only assist students with their technical studies, but also stimulate interest to enquire into some of the newer processes and techniques that are appearing in the field of engineering manufacture.

I would like to thank the various organisations for permitting me to use illustrations of their equipment, individual acknowledgments having been made at the relevant figures. Also a special thanks to Mrs. Rita Elley who kindly found time to type the manuscript ready for publication.

Finally, the introduction of TEC courses has meant a completely new approach to technical education at technician level. With regard to this I would welcome any comments or suggestions which could be considered for any future editions of this book.

P.J.H.

Contents

1 WELDING

Introduction 1. Welding processes 1. Types of fusion-welded joint 2. Metallurgical aspects of fusion welding 3. Gas welding 3. Oxy-acetylene cutting 6. Gas welding safety 7. Manual metal arc welding 7. Submerged arc welding 9. Shielded arc welding 10. Arc welding safety 11. Defects in fusion welds 11. Electrical resistance welding 12. Spot welding 12. Seam welding 12. Projection welding 13. Flash butt welding 13. Friction welding 14. Welding design considerations 15. Weld testing and inspection 16. Summary 19. Questions 20.

2 CASTING AND POWDER METALLURGY

Introduction 22. Shell moulding 22. Investment moulding 23. Diecasting 26. Characteristics of different casting processes 29. Powder metallurgy 30. Summary 31. Questions 32.

3 MEASUREMENT

Introduction 33. Comparators 33. Optical projection 38. Angle measurement 41. Surface finish measurement 46. Alignment testing of machine tools 50. Tests for centre lathes 50. Tests for horizontal milling machines 52. Tests for drilling machines 53. Limits and fits 53. Classification of fits 54. Limits-and-fits systems 56. Limit gauging 58. Summary 60. Questions 61.

4 MACHINING

Introduction 63. Single-point tools 63. Tool-force measurement 64. Power consumption 64. Metal removal rates 65. Tool life 65. Single-point tool construction 67. Milling 69. Types of cutter 69. Cutting techniques 73. Straddle and gang milling 74. Cutting speeds and feeds 74. Milling machine accessories 75. Helical milling 80. Grinding 81. Grinding wheel specification 82. Grinding wheel shapes 83. Grinding wheel speeds 84. Grinding wheel mounting 84. Balancing and trueing of grinding wheels 85. Grinding wheel dressing 86. Wheel forming 87. Characteristics of grinding 87. Grinding fluids 89. Grinding faults 89. Cutter grinding 90. Grinding safety 91. Summary 92. Questions 92.

5 CAPSTAN AND TURRET LATHES

Introduction 95. General machine design 95. Work holding devices 96. Standard tooling 97. Process planning 100. Machining costs 103. Summary 104. Questions 104.

1 Welding

INTRODUCTION   The term *welding* is used to describe a process for permanently joining metal by the application of intense localised heat. Unlike joints produced by soldering or brazing, a welded joint may be considered as a continuation of the material being joined and therefore having similar characteristics. This then makes welding suitable for structural applications where strong joints are required.

The origins of welding have been traced back as far as 3000 BC where simple forge welding was practised by ancient Roman and Egyptian civilisations. It was not however until the latter part of the 19th century that a serious effort was made to investigate welding as a commercial proposition. This resulted in a weld being produced by using the heat from an electric arc struck between a carbon electrode and the work, and thus formed the basis for modern electric arc welding in a variety of forms. Later, at the turn of the 20th century gas welding was invented whereby it was found that by mixing a suitable fuel gas with oxygen a high-temperature flame, sufficient for many welding applications, could be produced. It is often thought that gas welding was the forerunner of all welding processes, despite the fact that acetylene (the fuel gas normally used in gas welding) was discovered as long ago as 1836.

From these early beginnings welding has developed into a rapidly-expanding technology replacing some of the more traditional methods of joining, such as riveting. This development is still continuing mainly due to the introduction of new technologies and more especially due to the wide applications of automatic welding in large volume production.

There are many different types of welding processes and some of the more important ones are described in this chapter.

WELDING PROCESSES   All welding processes may be classified as being either *fusion* or *pressure* processes. In fusion welding a filler metal in the form of a rod or wire is heated together with the work. Sufficient heat is provided whereby both filler metal and the parent metal at the joint will melt, thus causing them to fuse together. Heat for fusion welding is obtained either by means of an electric arc or a flame from a blowpipe. In pressure welding no filler metal is used, the weld is produced by a combination of pressure, heat and time. Heat for pressure welding is usually obtained by electrical resistance.

Specific welding processes related to fusion and pressure

welding are shown in the following chart:

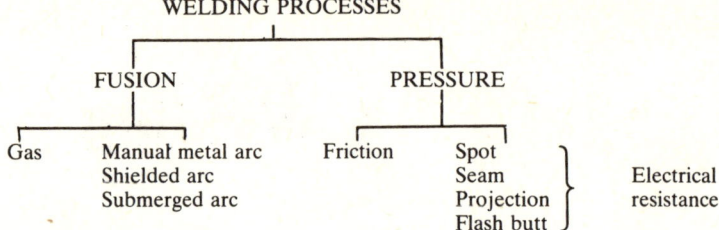

TYPES OF FUSION-WELDED JOINT

Joints for fusion welding consist basically of two types, these being *butt welds* and *fillet welds*.

Butt welds

These are used for joining plates edge to edge. A typical joint is illustrated in Figure 1.1 together with the terminology commonly associated with butt welds.

The type of joint preparation required depends mainly on the thickness of the materials. For material up to about 3 mm the edges may be flanged and fused together without the need for a separate filler metal, while for plates up to about 6 mm the edges may be left square. Plates thicker than 6 mm must have some form of edge preparation in order to achieve adequate penetration of the filler metal. Various forms of edge preparation are shown in Figure 1.2. The gap between plates is of special importance since this will affect the amount of penetration: with too small a gap the penetration will be insufficient; if too large, the molten weld pool will fall through and welding will be difficult to control. On very thick plate a U preparation is generally used instead of a V, since the amount of filler metal required is less. Although this must be considered

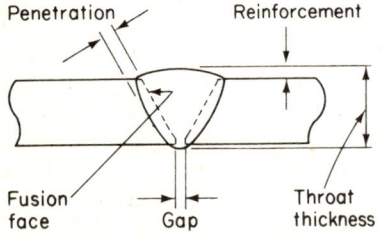

Figure 1.1 Butt weld terminology

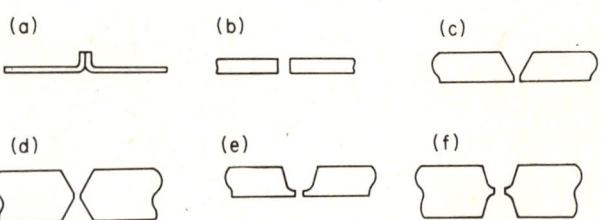

Figure 1.2 Butt weld joint preparations, (a) flanged, (b) square, (c) single V, (d) double V, (e) single U, (f) double U

against the extra cost of producing a U preparation as this must be machined, while a V preparation may be produced by flame cutting. For the exact dimensions of edge preparations reference should be made to BS 1856.

Fillet welds

A fillet weld is one where the two fusion faces are either inclined or at 90° to each other, as shown in Figure 1.3. Since edge preparation is not normally required for fillet welds these are often cheaper to produce than butt welds. A fillet weld may be one of two types, *double fillet* and *single fillet*, and although the former has sufficient strength for most applications, the

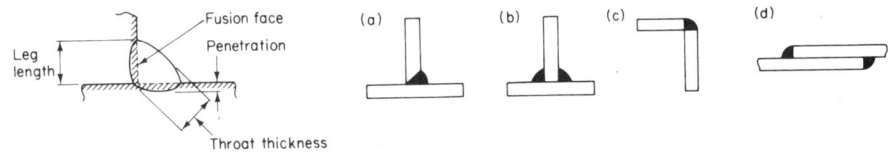

Figure 1.3 Fillet welds, (a) single, (b) double, (c) corner, (d) overlapped

single fillet is stronger since fusion takes place throughout the plate thickness.

METALLURGICAL ASPECTS OF FUSION WELDING

All fusion-welding processes involve melting, cooling and solidification of a filler metal. This means that the resultant weld may be likened to and possess characteristics similar to those of a miniature casting. If a weld is sectioned and prepared for examination then a structure similar to that shown in Figure 1.4(a) will be revealed. When a weld begins to solidify, it does so in the direction of cooling, this being towards the centre and producing long elongated crystals, known as *columnar crystals*, in the region of the fusion zone. The inner part of the weld will cool more uniformly resulting in an enlarged but regular crystal structure. Since the surface of the weld is in contact with air this will give rise to faster cooling producing a small and slightly chilled crystal structure. The parent metal will also undergo a structural change in the form of grain growth and this will be pronounced next to the weld metal in a region known as the *heat-affected zone*.

In gas welding the heat-affected zone tends to be wider, since the heat is concentrated for a longer period compared with arc-welding processes. Due to cast structures possessing relatively poor mechanical properties, fabricated parts are often given a post-welding heat treatment such as annealing to refine the weld and heat-affected zone. The weld shown in Figure 1.4(a) would have been accomplished in one pass, but where thick plate is to be joined several passes will be necessary. This has the advantage that the heat from subsequent passes will refine the crystal structure of the underlying previous weld (Figure 4.1(b)).

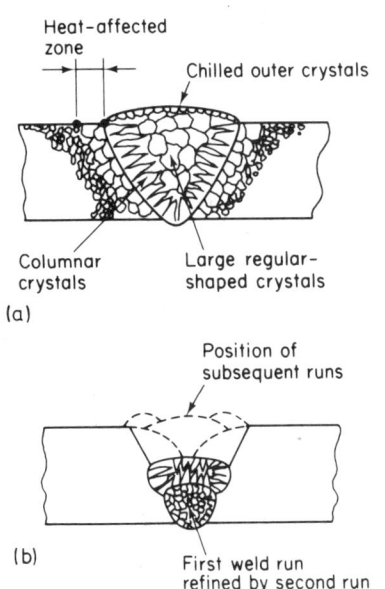

Figure 1.4 Weld metal structures

GAS WELDING

Oxy-acetylene welding is the most common gas-welding process. The required heat is obtained by mixing acetylene (fuel gas) with oxygen in a suitable blowpipe which after igniting provides a high temperature flame (approximately 3200°C). These two gases are stored in compressed form in two drawn steel cylinders, each cylinder being connected to the blowpipe by flexible hoses (Figure 1.5). Each cylinder is also fitted with a regulator to ensure that the gases enter the blowpipe at the correct pressure. The exact pressure used will be dependent on the blowpipe nozzle size and thickness of plate to be welded. For safety reasons welding gas cylinders are identified by a colour code, black for oxygen and maroon for acetylene. Furthermore, when acetylene is stored under pressure it may become unstable at elevated temperatures, which means that if an explosion is to be avoided then certain safety precautions must be built into the cylinder. This is achieved by packing the cylinder with a material such as asbestos or charcoal soaked in acetone to form a porous spongy mass, thus preventing the acetylene from separating explosively into its chemical constituents. Acetone is used because it is capable of absorbing approximately 25 times its own volume of acetylene per atmosphere of pressure ($0.1\,MN/m^2$). As an extra safety precaution acetylene cylinders are fitted with a fusible plug consisting of a low melting point alloy/which, after melting, will release any excess pressure.

The welding blowpipe

The function of the blowpipe is to enable the gases to be mixed in the correct proportion to produce the required flame. There are two basic types of blowpipe and these are designed to be used either with high- or low-pressure acetylene. The *high-pressure blowpipe* is the one most commonly used and is designed to permit both oxygen and acetylene to enter a mixing device in approximately equal volumes. This gas mixture is then directed to the nozzle where combustion takes place. The *low-pressure blowpipe* contains an injector through which passes oxygen at high pressure. Low-pressure acetylene is then drawn into this oxygen stream to obtain the correct mixture before combustion. Low-pressure blowpipes are seldom used with acetylene cylinders, since they are designed for use where acetylene is produced from a generating plant.

It is essential that all blowpipes are fitted with a 'flashback' arrestor. Flashback is caused by a flame being produced due to backfire at the hose connection of the blowpipe. This is a dangerous situation, not only from the point of view of damaging the equipment but may also be the cause of a possible explosion. To bring the blowpipe into operation, the acetylene is first turned on and ignited. The oxygen valve is then gradually opened and regulated until the required flame is produced.

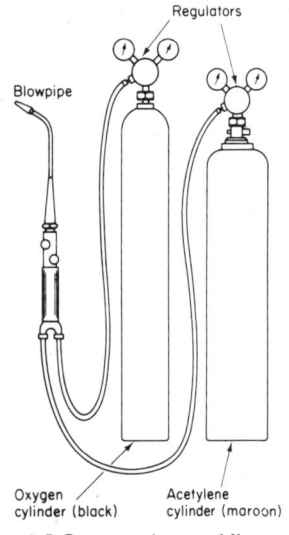

Figure 1.5 Oxy-acetylene welding equipment

Flame setting

It is important in all gas welding operations that the gases should be mixed in the correct proportions so as to produce a flame most suitable for the material to be welded. One of three types of flame may be used, these being *neutral*, *oxidising* and *carburising flames* (Figure 1.6). A neutral flame, the most common type, is one where equal amounts of oxygen and acetylene are burnt together and is suitable for welding ferrous materials, copper and aluminium alloys. The oxidising flame occurs when excess oxygen is used and is characterised by a small, brilliant white cone surrounded by a ragged bluish-white envelope. This flame is used mainly for welding brass. The carburising flame is the result of using excess acetylene and consists of a brilliant white cone surrounded by an acetylene feather, having a length approximately double that of the inner cone.

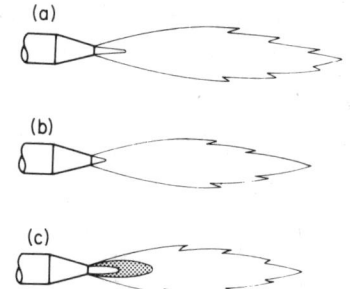

Figure 1.6 Oxy-acetylene flame settings, (a) neutral, (b) oxidising, (c) carburising

Gas welding technique

There are two principal methods of gas welding and these are

(i) Leftward welding (ii) Rightward welding.

Both techniques are illustrated in Figure 1.7.

In leftward welding the blowpipe and filler rod are moved from right to left, the filler rod being moved in a straight line while the blowpipe is moved backwards and forwards as the weld proceeds. It has been found by experience that the leftward method is suitable for welding plates up to about 5 mm thick. For thicker material the rightward technique is to be preferred. This time the blowpipe is moved from left to right in a straight line, while the filler rod is moved in a series of loops along the weld preparation.

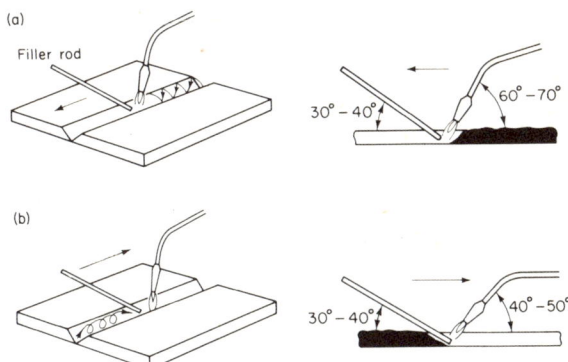

Figure 1.7 Welding techniques, (a) leftward, (b) rightward

Compared with leftward welding the rightward method has the following advantages:

(a) Since the flame is always directed towards the solidified weld, better mechanical properties are obtained due to the annealing effect.
(b) As the blowpipe moves in a straight line, very little agitation is produced, which could otherwise lead to excessive oxidation within the molten weld pool.
(c) Due to the better accessibility of the flame within the edge preparation smaller V angles are permissible. This also has the advantage that less filler metal is required thereby keeping gas consumption, welding time and hence cost to a minimum.
(d) Better vision for the operator.

Fluxes The majority of metals when heated will react with oxygen from the atmosphere to form an oxide on the metal surface. If precautions are not taken, this oxidation will be excessive during welding operations since the absorption of oxygen into the molten weld pool will result in a poor weld. So that oxidation is reduced a flux must be applied which should also have the ability of removing any oxides that are formed. The reader will no doubt be familiar with the application of fluxes during soldering operations.

When welding steel no flux is required since the oxide formed is lighter than the parent metal and therefore removes itself by floating to the surface of the weld in the form of a scale. This scale is easily removed when welding is complete. Metals such as cast iron, brass, bronze and aluminium alloys all require a suitable flux. Fluxes are usually applied in the form of a paste, although in some cases a dry powder may be used to advantage. It is usual to dip the filler rod into the flux to afford better protection as welding proceeds. Pre-coated filler rods are also available so that flux need be applied only to the parent metal.

OXY-ACETYLENE CUTTING

The principle of this process is that if a ferrous metal is heated and then exposed to an atmosphere of oxygen, then rapid oxidation will take place and this will cause the metal to break down. The process is therefore one of disintegration rather than actual cutting.

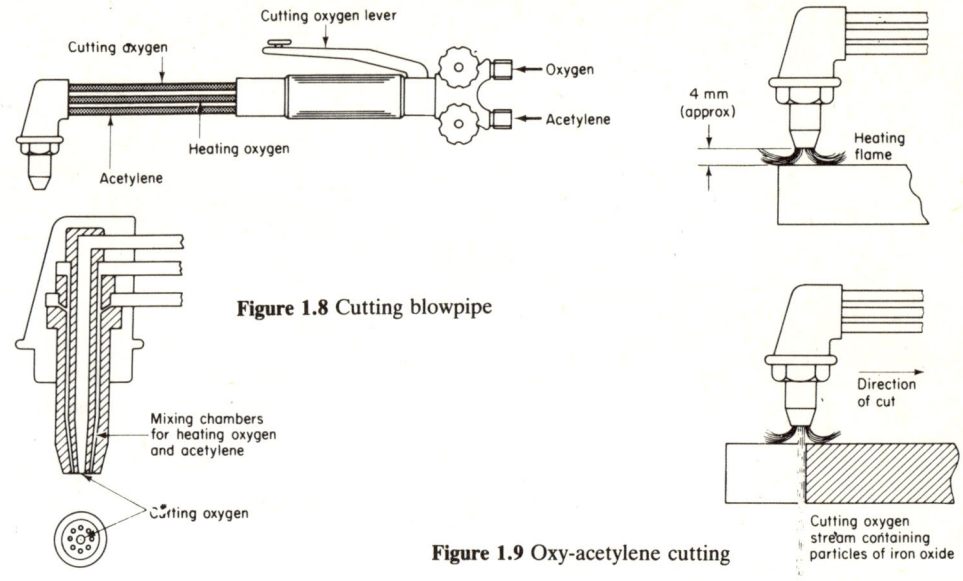

Figure 1.8 Cutting blowpipe

Figure 1.9 Oxy-acetylene cutting

The equipment used is similar to that used for gas welding, the main difference being in the design of the blowpipe. The *cutting blowpipe*, as it is more usually called, supplies oxygen and acetylene to provide a preheat flame and is further fitted with a lever to provide the additional oxygen for the cutting process. In the end of the blowpipe nozzle there is a central orifice to permit a flow of oxygen and a series of holes around this orifice to permit the flow of oxygen and acetylene mixture for the pre-heat flame. For most oxy-acetylene cutting processes a neutral flame is used, although for cast iron a carburising flame is generally used.

A typical cutting blowpipe is shown in Figure 1.8 together with a section through the cutting nozzle.

In operation the blowpipe is held about 4 mm above the surface of the work and pre-heated to a temperature of about 900°C. The oxygen is then turned on and the blowpipe moved at a uniform speed along the proposed cutting path, still maintaining a gap of about 4 mm between the nozzle and the work. The correct cutting speed is largely a matter of experience, if correct then edges should be sharp and smooth.

Gas cutting has a special place in the fabrication industry especially where thick plate is required, e.g. as those found in boiler and ship construction. The process is not confined to manual operation only, but lends itself to automatic control also, especially where complex profiles are required.

Gouging This is a technique whereby bulk material is removed or where a groove is required. The equipment used is the same as for flame cutting with the exception that specially-shaped nozzles are fitted to the blowpipe. Gouging is a useful process for preparing the edges of thick plate prior to welding. The process is illustrated in Figure 1.10.

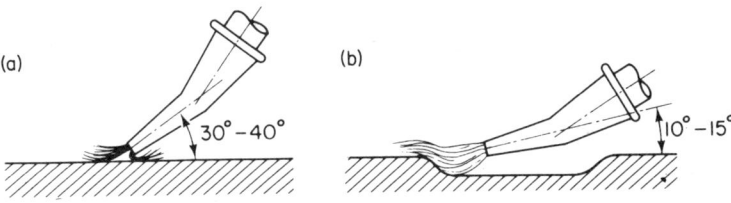

Figure 1.10 Nozzle positions for gouging, (a) for pre-heating, (b) for gouging

GAS WELDING SAFETY

The safety precautions relating to the storage of pressurised acetylene and the colour coding of gas cylinders have already been stated. As a further aid to equipment safety, so as to avoid the accidental mixing of gases, screwed hose and cylinder fittings are threaded with opposite hand threads. Oxygen fittings use right-handed threads, while acetylene fittings use left-handed threads. Furthermore, oxygen fittings must never be greased or oiled since these lubricants will react with oxygen with the risk of causing an explosion.

With regards to the welder's personal safety the following precautions should be observed:

(a) Due to the intense glare and radiation given off from the oxy-acetylene flame, operators must always protect their eyes by using approved goggles. A wide range of goggles is available having interchangeable filters to meet the requirements of different welding conditions.
(b) Adequate protective clothing must be worn. This usually takes the form of a leather apron and the wearing of leather or asbestos gauntlet-type gloves.
(c) When welding in confined spaces it is essential that an efficient ventilation system be provided. The fumes given off by welding operations can cause permanent damage to the respiratory system.

MANUAL METAL ARC WELDING

The principle of metal arc welding relies on a low-voltage, high-current circuit being interrupted by a small gap across which a high temperature arc will be produced. The temperature of this arc is in the order of 3500°C and provides sufficient heat for the welding of most metals. During welding this arc is maintained between a consumable metal electrode (filler) and the work.

Early arc welding processes used a bare metal electrode which had the disadvantage that the molten weld pool was exposed to the atmosphere, thus causing oxidation. Nowadays, flux-coated electrodes are used, which on melting form a protective gas shield around the electrode tip and molten weld pool. On cooling, the residue of this flux solidifies to form a slag on the surface of the weld, which is subsequently chipped away. The essential features of metal arc welding are shown in Figure 1.11.

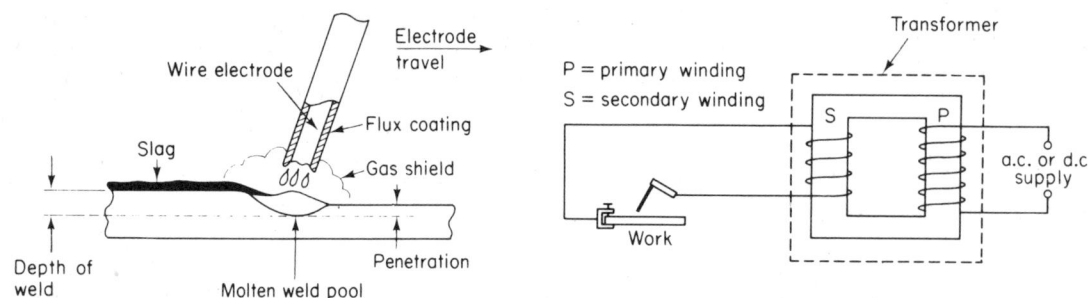

Figure 1.11 Manual metal arc welding

The voltage required to produce the arc is in the order of 60 to 100 V but, once started, the voltage required to maintain the arc is reduced to between 25 and 40 V. The power source used may be operated from either an a.c. or d.c. supply. When using a.c. supplies the polarity of the work and electrode is unimportant, i.e. it does not matter if the work and electrode are made positive or negative. However, with d.c. operation the polarity *is* important and the electrode manufacturer's instructions should be adhered to. For most steel welding applications alternating current is normally used, mainly due to the simplicity and relative cheapness of a.c. transformers. Equipment using direct current is more expensive since a transformer rectifier is required, but it is generally preferred for welding sheet metal, non-ferrous metals and stainless steels. The use of direct current has the advantage that it allows site welding to be undertaken where the power required is produced from a portable engine-driven generator.

Striking and maintaining the arc

Starting or *striking* the arc is one of the first aspects of metal arc welding that the trainee welder has to master. The simplest method of achieving this is to brush the electrode tip across the work surface and withdraw it until a gap of about 3 mm is produced. Alternatively, the electrode tip may be lightly tapped on the surface, so that a rebound equivalent to the arc gap is obtained. Once the arc has been struck it must be maintained at a constant length if a sound weld is to be produced. It follows then, that as welding proceeds, the welder must move the electrodes towards the work and control its speed of travel to give the required rate of weld deposition.

Electrodes

Electrodes used for metal arc welding basically consist of a metal core of a composition similar to that of the metal to be welded and surrounded by a flux coating. They are available in different diameters and lengths, the type selected depending mainly on the thickness of the parent metal and the welding current to be used. The diameter is of special significance, since this will have an effect on the amount of penetration achieved.

The type of flux coating used will depend on the weld metal composition. It should however be pointed out that the slag produced may chemically react with the underlying weld. The slag produced by aluminium fluxes is a common example and must therefore be removed by washing and brushing using hot water. An important characteristic of all fluxes is that the slag produced by them must have a lower relative density than the weld metal so that they float to the weld surface. This avoids the possibility of any trapped slag inclusions. The purposes of the electrode coating may be summarised as follows:

(a) To provide a gaseous shield so as to prevent oxidation to the molten weld pool.
(b) The combustion created between the electrode and work may cause a partial loss of certain elements such as carbon manganese nickel, etc. This may be overcome by including these elements in the flux coating to compensate for the loss.
(c) The solidified flux (slag) retards cooling of the weld, thereby refining its structure.
(d) To help stabilise the arc, especially when using alternating current.

Finally, of all the fusion welding processes, manual metal arc welding is perhaps the most widely used on account of the relative simplicity of the equipment and the ease with which the process may be performed. Over 75% of all fabrications in steel use the manual metal arc process.

SUBMERGED ARC WELDING

This method of arc welding is often a mechanised process whereby the arc and molten weld pool are entirely submerged in a layer of granular flux (Figure 1.12). The electrode used takes the form of a bare wire which is fed automatically from a reel through a feed nozzle. The flux is applied through a feed tube, just ahead of the electrode and takes its supply from a feed hopper. When in operation the lower layer of flux melts and provides a conductive path for the welding current. Since the upper portion of the flux is not affected this may be re-used.

Compared with other arc welding processes, submerged arc offers the following advantages:

(a) Due to the arc being completely covered by flux, high currents of up to 4000 amps may be used. This permits the welding of thicker plate, compared with that used for other arc welding processes.
(b) High welding speeds are possible, e.g. for welding 12-mm steel plate 10 mm/s may be achieved. The welding speed used will be dependent upon the plate thickness, voltage and current used and nature of the joint preparation.
(c) Since the arc is entirely submerged there is no visible arc glare, smoke or flash, thereby eliminating the need for operators to wear protective shields or helmets.

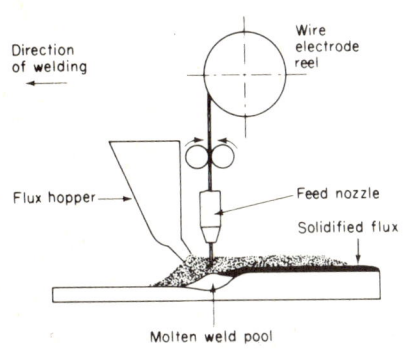

Figure 1.12 Submerged arc welding

The capital cost of submerged arc welding equipment is relatively high, but since high welding speeds are possible, welding time (and hence cost) is considerably less compared with metal arc processes. The main application of the process is for welding long continuous lengths, especially those of a repetitive nature. Although confined mainly to welding in the horizontal plane, vertical and inclined welding may be performed with minor modifications to the equipment.

SHIELDED ARC WELDING

The term *shielded arc welding* refers to a group of welding processes whereby an inert protective gas shield, usually argon, is introduced around the region of the arc and molten weld pool. Although this gas shield is primarily used to prevent oxidation of the deposited weld metal, other gases may be mixed with it for specific purposes. For example, a small percentage of oxygen added to argon improves the action of the arc and general welding characteristics of certain stainless steels.

The two most common shielded arc processes are

(i) Tungsten inert gas (TIG) (ii) Metal inert gas (MIG).

Tungsten inert gas In TIG welding the arc is produced between a tungsten non-consumable electrode and the work, the filler metal being supplied in the form of a separate rod. The protective shielding gas is delivered through the electrode holder, as shown in

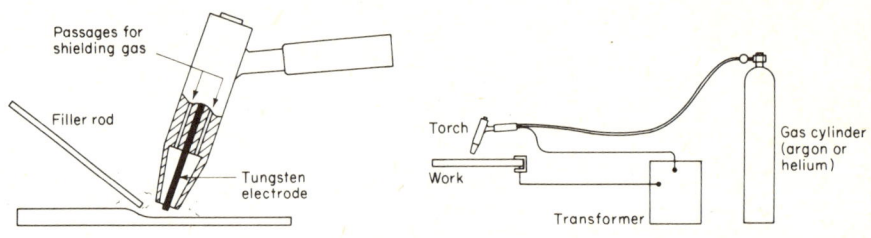

Figure 1.13 Tungsten inert gas (TIG) welding

Figure 1.13. The use of TIG is confined mainly to the welding of relatively thin materials of up to about 7 mm. Although a wide range of materials may be welded the process finds special application for welding aluminium- and magnesium-based alloys.

It has already been stated that argon is the usual shielding gas. In some cases helium may be used which is found beneficial for welding thicker sections since light arc voltages are possible with this gas. The main disadvantage of helium is its high cost. In general the use of TIG welding requires a high degree of operator skill.

Metal inert gas In MIG welding (Figure 1.14) the arc is produced between a consumable electrode in the form of a wire fed through a holder from a reel. The shielding gas is delivered through the electrode

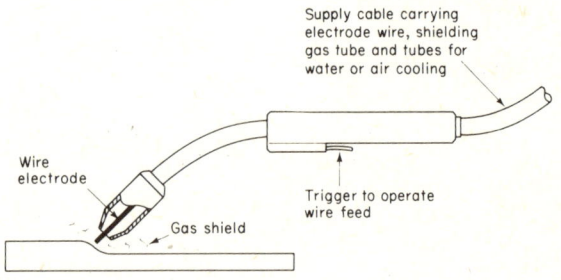

Figure 1.14 Metal inert gas (MIG) welding

nozzle. Since the wire electrode is small in diameter compared with other welding electrodes the rate of consumption must be high in order to obtain the maximum amount of deposited weld metal. This necessitates a mechanical feed through the welding gun, up to about 10 m/min. The exact wire speed will be governed by the welding current and the type of shielding gas used.

The shielding gases used are similar to those for TIG welding, although carbon dioxide (CO_2) and argon/CO_2 mixtures are often used for welding various types of carbon steels. The main advantage of using CO_2 is that it is considerably cheaper than argon.

To a limited extent the use of MIG welding has replaced TIG mainly for welding thick sections. Compared with TIG very little welding skill is required in order to obtain a satisfactory weld.

ARC WELDING SAFETY

With the exception of submerged arc, all arc welding processes are identified with an intense arc glare, consisting of ultra-violet radiation. In addition to hot particles being emitted from the arc, this radiation will cause discomfort to the welder together with the risk of long-term health damage if precautions are not observed. No part of the welder's skin must be exposed to the rays from the arc since burning will result. The type of protective clothing worn usually takes the form of leather aprons, jackets and gloves, and in some cases leather spats to afford protection to the feet and ankles. Special attention must be given to the protection of the face and eyes. An irritating condition that welders sometimes suffer from is known as 'arc eye'. Protective face shields or helmets are used incorporating specially-prepared filters which are interchangeable to suit different welding conditions. On no account must oxy-acetylene goggles be worn since these will give no protection to the face. Other personnel working near to arc welding processes should also be protected, and this is best achieved by erecting screens to shield off the arc rays. Further, it is also desirable that walls and ceilings of welding shops be painted a suitable colour so as to absorb and not reflect ultra-violet rays.

DEFECTS IN FUSION WELDS

The ideal weld should be such that adequate fusion exists between the filler metal and edge preparation together with good penetration. It is also desirable that a slight reinforcement exists above the two welded parts. There are, however, several defects that will not always be apparent on the weld surface. These include

(i) Insufficient fusion (iii) Porosity
(ii) Insufficient penetration (iv) Cracking.

Insufficient fusion is usually caused by insufficient heat and too fast a travel of blowpipe or electrode. Insufficient penetration is often caused by incorrect edge preparation or using wrong diameter of filler rod/electrode. Porosity results in a number of small holes throughout the weld, caused by gases being trapped within the molten weld pool. Cracking may occur due to incorrect welding technique or using a filler metal having a different rate of contraction compared with that of the parent metal.

ELECTRICAL RESISTANCE WELDING

Welding processes in this group acquire the necessary heat due to the electrical resistance set up through the joint material. When the correct temperature is reached pressure is then applied for a prescribed period of time to complete the weld. No additional filler metal is required.

Spot welding

The application of spot welding is confined mainly to the joining of thin sheet material using overlapped joints in both ferrous and non-ferrous metals. The maximum thickness that can be accommodated by the process seldom exceeds 8 mm. In operation the two surfaces to be joined are held together and under pressure between two copper electrodes. While this pressure is applied a high current is passed from one electrode to the other and through the work for a specified period of time. The combination of pressure and heating effect due to the electrical resistance of the work material causes the two surfaces to join together locally, as shown in Figure 1.15. An important

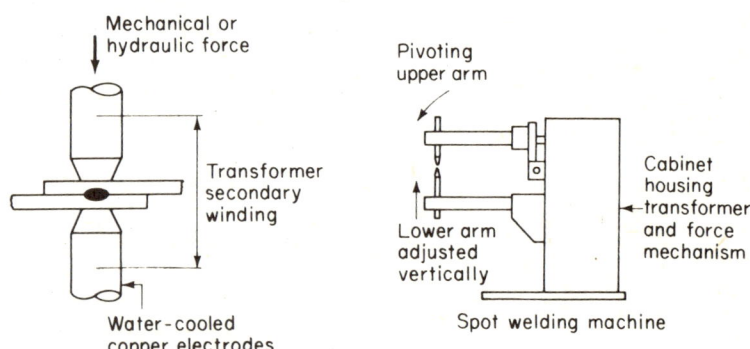

Figure 1.15 Spot welding

consideration of the process is that the correct electrode diameter should be chosen which will be related to the material thickness to be joined. Electrodes, especially on large machines, are water-cooled.

Seam welding

This process is an adaptation of spot welding, designed to produce a series of continuous, slightly overlapping spot welds in thin sheet metal. The machines used are similar to those used for spot welding, except that the vertical-rod-type electrodes are replaced by copper roller electrodes. In operation the work is fed between the two rollers and the electric current pulsed at regular intervals depending on the feed rate of the work. This produces the characteristic joint as shown in Figure 1.16(a).

Seam welding is generally used where a water- or gas-tight joint is required, e.g. on liquid storage tanks and automobile silencers. Where a water- or gas-tight joint is not required the

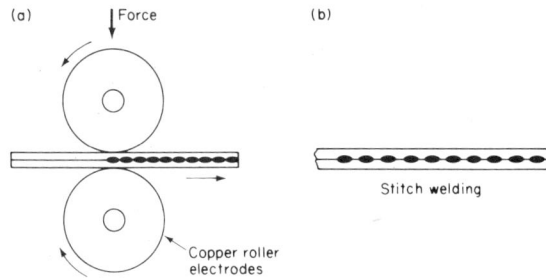

Figure 1.16 Seam welding

current pulse rate may be changed to produce an interrupted seam. This modification to the process is often referred to as *stitch welding* (Figure 1.16(b)).

Projection welding This process illustrated in Figure 1.17(a), is similar to spot welding, the principal difference being that the heat generated is concentrated around projections formed on one part of the workpiece, instead of on electrodes as in spot welding. In operation the parts to be joined are placed between two copper platens (electrodes) through which the welding current is passed via the projections. On welding temperature being reached, pressure is applied, and this results in collapse of the projections which form into welds similar to those obtained by spot welding.

The application of projection welding is confined mainly to components produced by press tools, since it is relatively easy to form the required projections during presswork operations. In some cases, projections may be formed by machining, as for example the welding of a stud to a thin metal plate, as shown in Figure 1.17(b).

The advantages of projection welding may be summarised as follows:

(a) Multiple spot welds may be made simultaneously, thereby reducing the cost per component.
(b) Spacing of welds may be reduced. In spot welding, if welds are made close together then a proportion of the welding current tends to shunt through the previous weld and cause a weak joint.
(c) Slightly less current required, thus promoting longer electrode life.

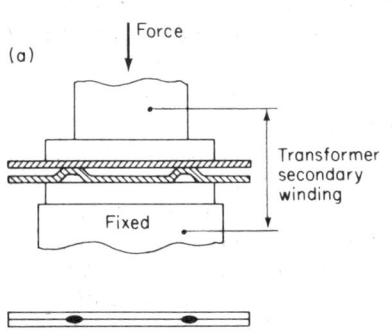

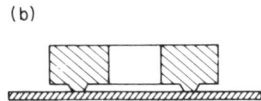

Figure 1.17 Projection welding

Flash butt welding This process finds application for butt-joining bars, tube and extruded sections. In operation the two parts to be joined are secured in clamps and the joint faces brought into light contact (Figure 1.18). When the electric current is applied arcing takes place between the two surfaces and this causes a characteristic flashing action. This arc is maintained by moving one joint face towards the other, during which time both faces increase in temperature until they become plastic. As soon as the correct temperature is reached the current is switched off and pressure rapidly applied to complete the weld. This pressure will also squeeze out from the joint faces any unwanted slag, oxides and overheated metal. Due to the upsetting action of the process,

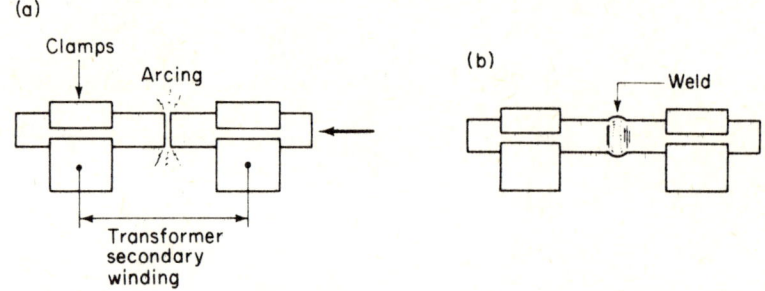

Figure 1.18 Flash butt welding

flash butt welded joints are always accompanied by a slight bulging around the weld. If troublesome or undesirable this may be machined off without causing any loss of strength to the weld. Flash butt welding is often used for joining angle and other sections such as bars and tubes.

FRICTION WELDING

Unlike the resistance welding processes so far discussed, this process uses the heat produced by friction and not that due to the heating effect of an electric current. The process involves the rotation at high speed of one of the two parts against the other which is firmly fixed. During this operation frictional heat generated by the pressure applied will cause the joint faces to become plastic. At this point the rotating member is quickly brought to rest as the pressure is increased so that the weld is produced by the two parts being forced together. The resultant joint is always characterised by an upset annulus around the weld which may be subsequently removed. The principle of friction welding is illustrated in Figure 1.19(a).

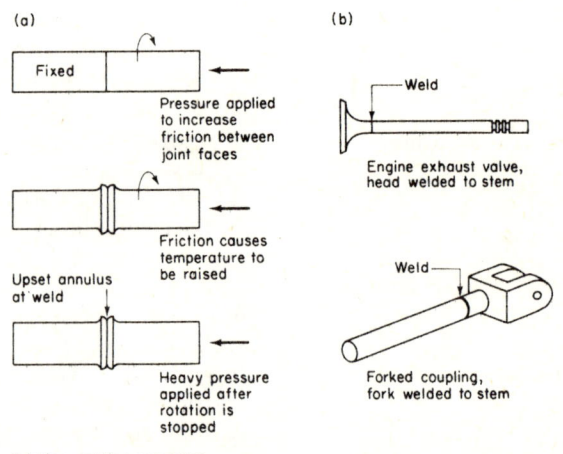

Figure 1.19 Friction welding

For the majority of ferrous materials no joint preparation is required since the rotating action is self-cleansing and any surface irregularities of the joint faces will be corrected during the welding cycle. For certain non-ferrous metals preparatory cleaning of the joint faces is important and these should be of machined quality. The application of friction welding is confined to parts where at least one part is of circular cross-section, i.e. tube or bar. Typical examples of friction welding are shown in Figure 1.19(b).

WELDING DESIGN CONSIDERATIONS

Modern welding methods enable designers to make use of a wide range of materials for fabricated structures. In many cases this has meant that, due either to a lack of production facilities or improved manufacturing cost, parts are made as fabricated assemblies instead of being produced by more traditional methods such as casting. It also means that the designer must understand how metals will react when subjected to welding operations, since the intense heat used can cause distortion.

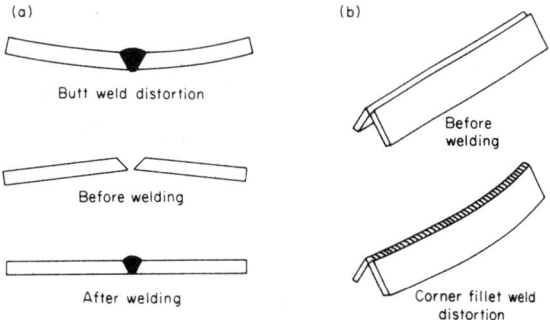

Figure 1.20 Weld distortion

Furthermore, it is important to ensure that parts are welded in the correct sequence, not only to minimise distortion but also to minimise the possibility of contraction stresses being set up. These are stresses caused by welded parts not being able to contract freely on cooling which in turn may cause distortion of fabricated assemblies. Examples of distortion are shown in Figure 1.20. The butt joint shown in Figure 1.20(a) may be given a double V preparation whereby the second weld run will pull out the distortion created by the first run. Alternatively, the plates may be laid at a slight angle so that after welding they will pull up horizontal.

Where a number of parts have to be welded together, especially where the operation is of a repetitive nature, welding fixtures are frequently used. Since these fixtures use clamps which rigidly hold the parts to be welded, it is advisable to make a series of short welds first (*tack welding* (Figure 1.22)) and then finish with full welds when the assembly is released from the fixture. This procedure will assist in keeping contraction stresses and distortion to a minimum. Where service requirements specify welded structures to be stress-free, a stress-relieving heat treatment may be carried out, usually within the temperature range of 250 to 300°C.

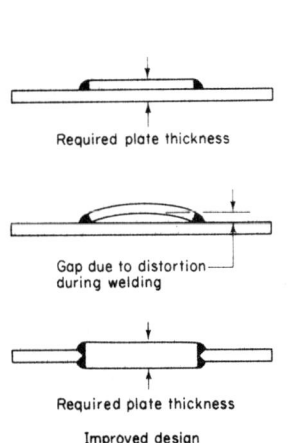

Figure 1.21 Method of increasing plate thickness in fabricated structures (Institution of Production Engineers)

Many fabricated structures require local thickening of a parent plate to form bosses, mounting pads, etc. The easiest solution might be to weld on a second plate, but this is regarded as bad practice since distortion as shown in Figure 1.21 will occur. The gap produced by this distortion is also a potential moisture trap which could act as a possible source of corrosion.

Where additional strength is required, stiffeners and gussets are frequently used. Although the positioning of these is governed by load distribution considerations it may be sufficient to attach them by tack welding. This will keep the heat from the welding process to a minimum and at the same time prove economic in the use of the filler metal. It should be appreciated, however, that for some fabrications which will be subjected to extreme loadings a full weld may be required. Gussets are often used to provide stiffness at the corners of welded structures. Here it is good practice to clear the corner of gusset plates so that the possibility of stress concentration is eliminated. The welding of stiffeners and gusset plates is illustrated in Figure 1.22.

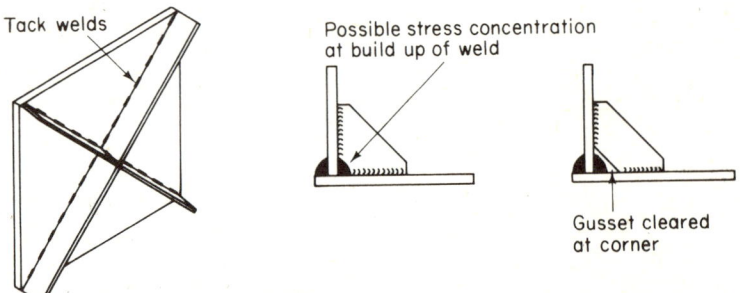

Figure 1.22 Use of stiffeners and gussets in fabricated structures (Institution of Production Engineers)

WELD TESTING AND INSPECTION

The simplest method of weld inspection is by visual examination. Although this test will not reveal any internal defects it may however be all that is required for certain classes of work. Where more information is required more positive tests may be employed which may be of either a destructive or non-destructive nature.

Destructive testing

Destructive testing includes tensile, impact and bend tests. *Tensile* and *impact testing* of welds is carried out in a similar fashion as for standard tensile and impact testing procedure. *Bend tests* are usually carried out to assess the soundness and ductility of a welded joint. The angle through which the test piece is bent must be pre-determined since this will be related to the mechanical properties of the material. A common form of bend test is to bend the test piece through 180°, as shown in Figure 1.23.

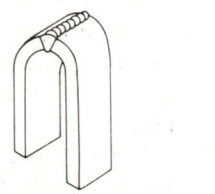

Figure 1.23 180° bend test

Non-destructive testing

The disadvantage of destructive testing methods, of course, is that the welded part is destroyed and therefore these tests are suitable only for weld testing on a sample basis. This restriction may be overcome by using various non-destructive testing methods based on different principles to suit different conditions. It should, however, be emphasised that the interpretation of the results requires a certain amount of skill and experience.

DYE PENETRANTS

The use of dye penetrants represents the most common and simplest method of non-destructive weld inspection and is used for the detection of surface cracks (Figure 1.24). The method consists of brushing or spraying the surface of the weld with a coloured, penetrating dye which is allowed to seep into any cracks present, any excess dye being wiped away using a suitable cleaning fluid. At this stage care should be taken not to draw out dye from the crack. A developer, usually in the form of a dry powder, is now applied to the surface. Any residual dye will then be drawn out into this developer causing it to spread slightly so that the presence of any surface cracks is rapidly detected. Differently coloured dyes are available so that the best contrast can be obtained against the surface under test.

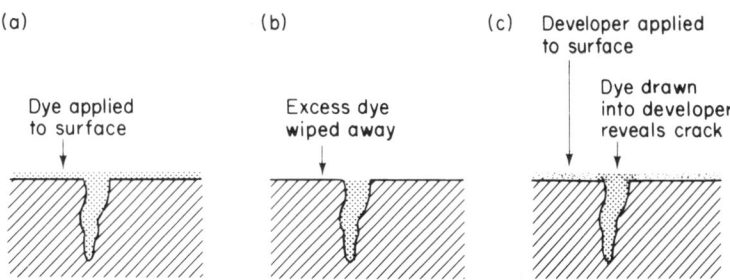

Figure 1.24 Crack detection using dye penetrants

Fluorescent dyes are also available which give a clearer indication when viewed under ultra-violet light.

Dye penetrants are suitable for use with all ferrous and non-ferrous metals. Their main advantage compared with other non-destructive testing methods is that the equipment is very simple and of low cost.

MAGNETIC CRACK DETECTION

This test is suitable only for ferro-magnetic materials. It uses electromagnetic induction to induce a magnetic field into the workpiece. Any discontinuity, e.g. a crack, will cause this field to be interrupted and in so doing will cause opposite poles to form at the edges of the crack. A liquid solution containing iron filings, sometimes known as a 'magnetic ink', is applied to the surface and the filings will be attracted to the crack where the poles have formed, as illustrated in Figure 1.25.

An important requirement of this test is that the weld surface should be reasonably smooth, otherwise the results obtained may be misleading. Submerged arc and shielded arc processes usually do not cause any problems, but welds produced by using flux-coated electrodes may require preparatory grinding before the test is performed.

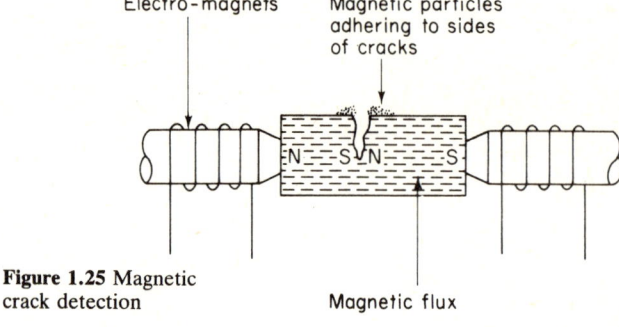

Figure 1.25 Magnetic crack detection

A further consideration is that the only surface cracks which will be revealed are those lying at approximately 90° to the direction of the magnetic flux.

RADIOGRAPHY

The non-destructive testing methods so far discussed will reveal irregularities only at the surface of the weld. For detecting internal defects such as blowholes, slag inclusions and porosity, radiographic methods using X-rays may be used.

The equipment consists basically of an evacuated glass tube into which are fitted a cathode and an anode, the cathode taking the form of a filament and the anode a platinum target. When a high voltage (approximately 120 000 V) is applied across the cathode and anode the filament becomes hot and emits negatively-charged particles known as *cathode rays*. On striking the target (anode) they are reflected to produce electromagnetic radiations of short wavelength which are more commonly known as X-rays.

These rays have the ability to penetrate and be absorbed by solid substances, the degree of absorption being dependent upon the density of the substance. If, for example, a blowhole or internal crack exists within the weld, then the amount of absorption will be less at these defects compared with the surrounding metal, since the weld density would have been locally reduced. This variation in absorption is detected by means of a photographic plate positioned behind the workpiece, as shown in Figure 1.26.

Another radiography technique involves the use of gamma rays which, compared with X-rays, are radiations of shorter wavelength. The main advantage of using gamma rays is that the equipment is much cheaper compared with that needed for X-rays, although the applications are not as great, especially where fine defects are to be detected.

Due to the dangerous nature of X-rays and gamma rays, personnel operating such equipment are required to take certain safety precautions. In addition to protective clothing, radiation monitors must also be worn. These consist of film badges which after exposure to radiation are developed to reveal the extent of the radiation dosage which must be within an acceptable level as specified by the Health and Safety Executive.

ULTRA-SONIC TESTING

The term *ultra-sonic* refers to a group of sound waves that are outside the scope of human audibility, but find a special use for

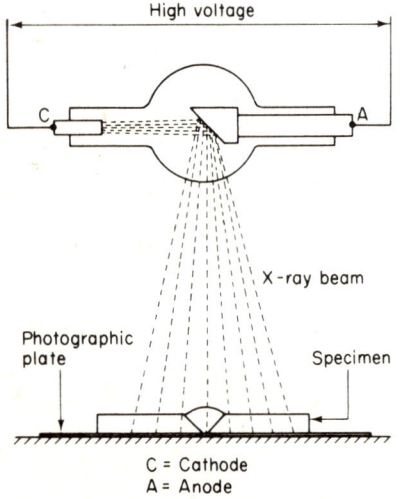

Figure 1.26 Radiographic inspection

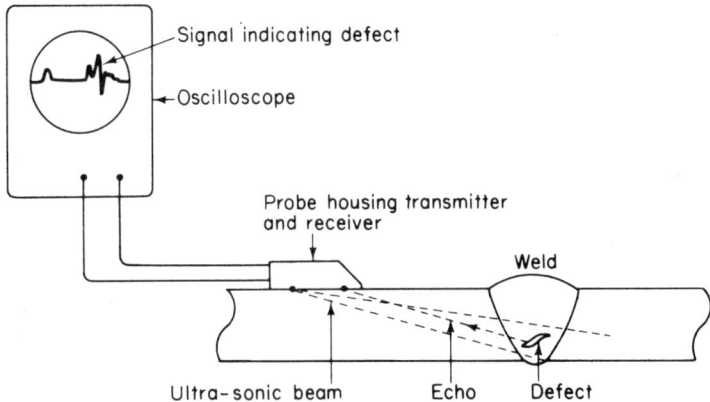

Figure 1.27 Ultra-sonic weld inspection

the detection of internal defects within solids (Figure 1.27). The principle of the test consists of transmitting ultrasonic waves from a probe through the work-piece and any defect existing within the weld will cause an echo to be produced, due to the partial reflection from the defect of the transmitted sound waves. This echo is detected by a receiver probe which converts the reflected sound waves into an electrical signal which may be displayed onto an oscilloscope screen. The two probes used must make contact with a smooth surface, via a coupling fluid which must be used since ultrasonic waves will not transmit sufficiently through air. Since the surface of a weld is never smooth the probes are usually positioned to one side and the transmitted signal projected at an angle.

For a full examination of the weld, length and depth scans are necessary. The length scan simply involves moving the probes along a line parallel to the weld. With the depth scan the probes must be moved at 90° to the weld, so that the transmitted beam will detect defects in the vertical plane. Just as with X-ray examination, skill is required in the interpretation of the results, and depending on the nature of the signal displayed an experienced inspector will be able to identify the nature and size of the defect, e.g., blowhole, crack, etc.

SUMMARY Welding processes are classified as being either fusion or pressure processes.

Fusion-welding processes include oxy-acetylene, manual metal arc, submerged arc, and shielded arc, and all these require the use of a separate filler metal in the form of a rod to complete the weld. By using a suitable blowpipe, oxy-acetylene equipment may be used for flame cutting and gouging.

Welds produced by the above fusion processes are known as either butt or fillet welds. *Butt welds* require the plate edges to be specially prepared, but *fillet welds* do not require any edge preparation.

Due to the intense glare from fusion-welding processes operators must ensure that their eyes are protected. Approved welding goggles are suitable for gas-welding processes, but face shields or helmets must be worn when using arc welding equipment. Additional protective clothing is also essential.

Pressure-welding processes include spot, seam, projection, flash butt and friction welding. No filler metal is required for these processes. Spot, seam and projection welding are carried out on thin sheet metal while flash butt and friction welding are used for producing butt welds on bar or metal of complex section.

Due to the heat created during welding, the work often distorts, and if possible this should be allowed for so as to avoid the scrapping of what is otherwise good work. Careful design of fabricated structures will also minimise the risk of distortion. Welds should be inspected for strength and soundness, the tests used being of either a destructive or non-destructive nature. *Destructive tests* include tensile, impact and bend tests. *Non-destructive tests* include the use of dye penetrants, magnetic crack detection, radiography and ultra-sonics.

QUESTIONS

(1) Explain clearly the difference between
    (i) Fusion welding (ii) Pressure welding.

(2) Name and describe, using sketches, the three types of flame used in oxy-acetylene welding. State under what conditions each flame would be used.

(3) Explain clearly the difference between rightward and leftward welding. State three advantages of the rightward method.

(4) Describe, using a sketch, the principle of manual metal arc welding. State clearly how protection to the molten weld pool is achieved during the process.

(5) Compare the use of a.c. and d.c. supplies used in manual metal arc welding.

(6) Describe, using a sketch, the principle of submerged arc welding. State two main advantages in using this process.

(7) With the aid of sketches explain the difference between the following shielded arc welding processes:
    (i) Tungsten inert gas (TIG)
    (ii) Metal inert gas (MIG).

(8) Outline the main areas of application for both TIG and MIG welding processes.

(9) Explain clearly the difference between a butt weld and a fillet weld. Sketch three types of joint preparation often used for butt welds.

(10) With reference to a welded joint explain what is meant by the following terms:
    (i) Heat affected zone
    (ii) Penetration
    (iii) Reinforcement.

(11) Name two possible defects that may occur in a weld and state their possible causes.

(12) Describe with the aid of a sketch the construction of an oxy-acetylene cutting blowpipe.

(13) With reference to oxy-acetylene cutting what is meant by the term *gouging*.

(14) State two reasons for the flux coating used on electrodes for manual metal arc welding.

(15) Explain clearly using sketches the difference between spot and seam welding.

(16) Explain why projection welding is sometimes used instead of spot welding. Sketch the type of preparation required for the projection welding of two thin metal plates.

(17) Using sketches describe the principle of operation of the following processes:

   (i) Flash butt welding (ii) Friction welding.

(18) Two 10 mm thick steel plates are to be butt welded together. Describe two possible methods that may be used to minimise distortion.

(19) Name three methods of non-destructive testing. Explain any one of your choice and state its main applications and limitations.

(20) Outline the safety precautions that should be observed when using fusion welding processes.

2 Casting and powder metallurgy

INTRODUCTION Casting in various forms represents one of the most important metal-shaping processes used in engineering manufacture.

Readers will already be familiar with sand casting from their studies at level 2. Although a versatile process and capable of producing castings of complex form in a wide range of metals, the dimensional accuracy and surface finish produced by sand casting are relatively poor. Furthermore, sand casting is not generally suited to large volume production, especially where castings are requried to have fine detail. To avoid these limitations, other casting processes, which also have lower manufacturing costs, have been developed. These include

(i) Shell moulding (ii) Investment moulding (iii) Diecasting.

In addition to casting processes, the shaping of components by the use of metal powders is also included in this chapter.

SHELL MOULDING This process may be regarded as a development of sand casting. Basically the process consists of making two consumable half-moulds or shells from a sand mixed with a suitable resin binder which has been cured so that the shells possess sufficient strength to withstand the weight of the cast metal.

Forming the shell To form the shell a metal half-pattern is first made, usually in steel or brass, and attached to a pattern plate. A pattern for the runner is also included on this plate. Draught angles of about 1 to 2° are provided on the pattern to facilitate stripping, together with stripper pins incorporated into the pattern plate design.

Partial curing The whole assembly is now heated to a temperature of about 250°C, which may be carried out in an oven or by electrical resistance heaters fitted within the pattern. Whichever method of heating is used, the pattern plate assembly is clamped to a dump box containing sand mixed with a thermo-setting resin. This dump box is now inverted so that the sand/resin mix falls onto the heated pattern which causes the resin to melt and form a bond for the sand. After about 10 to 25 seconds the dump box is rotated back leaving a thin (approximately 5 mm) partially-cured shell adhering to the pattern.

Final curing and pouring The pattern plate assembly together with the shell is now transferred to an oven where final curing is carried out within the temperature range of 300 to 400°C, for a period between one and five minutes. The actual time and temperature used will depend on the type of resin used. After curing, the shell is stripped from the pattern plate. Both shells are made in this fashion and the mould completed by joining the two halves together by using bolts, clamps or adhesives. The mould is now ready for pouring. Where hollow sections are required cores are included, and these are prepared in a fashion similar to sand casting. The stages in producing a shell mould are illustrated in Figure 2.1.

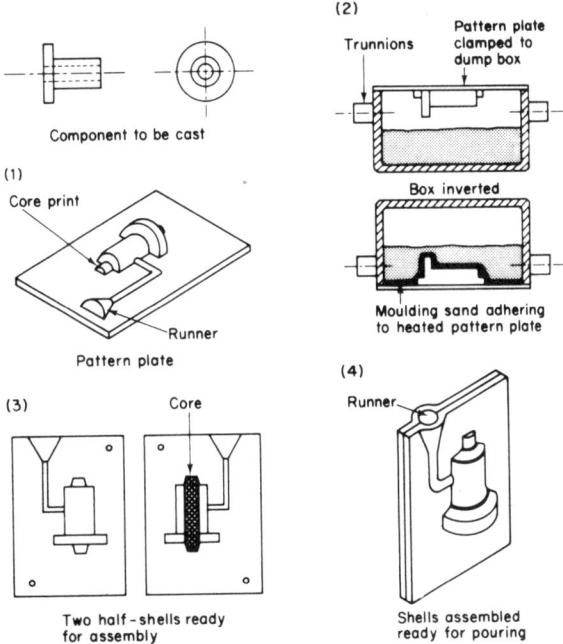

Figure 2.1 Stages in producing a shell moulding

Compared with sand casting, shell moulding offers the following advantages:

(a) Better dimensional accuracy, with a tolerance of ± 0.10 mm possible.
(b) Improved surface finish together with the reproduction of finer detail.
(c) The process lends itself to the use of unskilled or semi-skilled operatives.

The main disadvantage is the high cost of the patterns and moulding sand. However, since the process is semi-mechanised the cycle time involved in producing a shell mould is considerably less compared with producing a mould for sand casting. This therefore makes the process more suited to the production of castings in large quantities where the initial tooling cost can be offset.

INVESTMENT MOULDING This method of casting can claim to be as old as sand casting and was used by the ancients for producing articles requiring fine detail such as those found on sword hilts and jewellery, etc. Throughout the centuries the process was confined mainly to the production of bronze sculptures and is indeed a process still used today for this art form. It was not until the first quarter of this century that investment moulding was found to be suitable as an industrial process, especially where castings with a high dimensional accuracy and good surface finish are required.

Basically the process consists of making a wax pattern, which is invested with a refractory (heat-resistant) material to form the mould. When the investment has hardened the wax is melted out leaving a mould cavity which is then filled with molten metal. When the metal has solidified the refractory mould is broken away to reveal the casting.

The process may be divided into three distinct stages:

(i) Making the pattern (ii) Investing the pattern
(iii) Casting the metal.

Making the pattern To produce the wax pattern a split die is required which basically may be made in one of two ways, as follows:

(1) Where a long life is expected, dies may be made from metal; steel, brass and aluminium are commonly used. A reverse profile is machined into the metal with allowance being made for shrinkage. The degrees of skill and precision required are high, similar to those found in the manufacture of dies used for plastic moulding.
(2) Where the die life is not so important a cheaper die may be made from a low melting point alloy. The procedures illustrated in Figure 2.2. The first requirement is a master

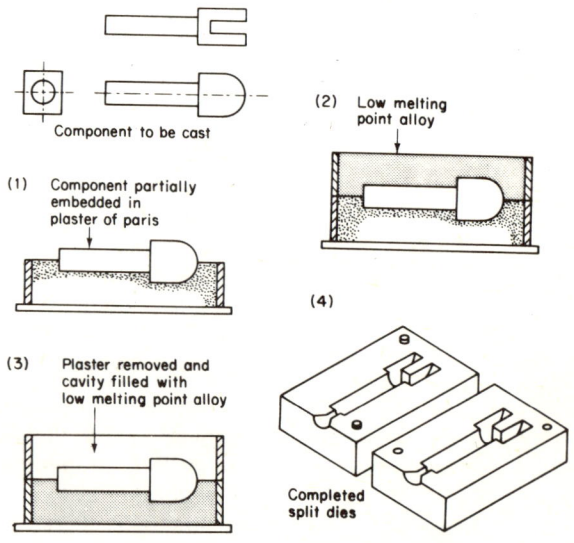

Figure 2.2 Stages in producing dies from a low melting point alloy

pattern made from steel or brass which is given a good surface finish and made slightly oversize to allow for shrinkage of the wax. This pattern is sunk into a bed of plaster of Paris to a depth corresponding to the parting line of the die. A steel flask positioned around the remaining half pattern is then filled with a low melting point (190°C) tin/bismuth alloy. When this has solidified the two halves are turned over and the plaster is removed to be replaced by the same low melting point alloy as before.

Whichever method of making the dies is used they are then assembled and injected under pressure with molten wax. On solidification the dies are split and the wax pattern removed. Where small castings are to be produced in large quantities, it is usual to attach the patterns to a wax spine so as to form a tree (Figure 2.3) prior to investment.

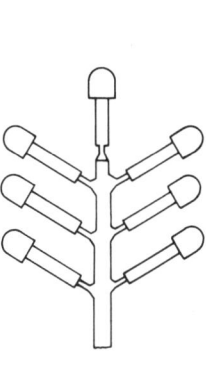

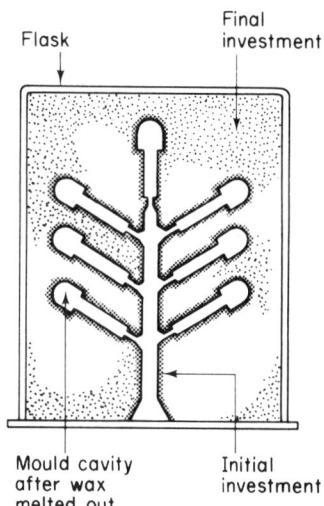

Figure 2.3 Wax patterns formed into a tree

Figure 2.4 Completed mould after investment

Investing the pattern   The refractory coating applied to the pattern to form the mould is known as *investment* and this is carried out in two stages. Initial investment consists of spraying with, or dipping the pattern into, a water mix solution consisting of sodium silicate, chromic oxide and zircon flour. Before this coating dries it is usually dusted with a coarser refractory material known as molochite to provide a key for the final investment. After drying, the coated pattern is surrounded by a metal flask which is then filled with a second investment material consisting of a solution of water mixed with fused alumina or fused fireclay. To ensure that this refractory is compacted around the first investment the flask is mechanically vibrated for a few minutes. The flask is then placed in a low-temperature furnace for initial hardening of the investment and also to melt out most of the wax, which may be re-used. This operation lasts for about eight hours at a temperature of about 95°C, the exact time and temperature being dependent on the type of wax used. The temperature is then raised to about 1000°C for final hardening of the mould and to melt out any remaining wax. The mould is now ready for pouring, as shown in Figure 2.4.

Casting the metal   While still hot the mould is clamped to an electrically-heated tilting furnace (Figure 2.5) containing an accurately-charged amount of molten metal. When at the correct temperature the furnace is inverted thus allowing the molten charge to flow into the mould cavities. To ensure that the metal flows to all parts of

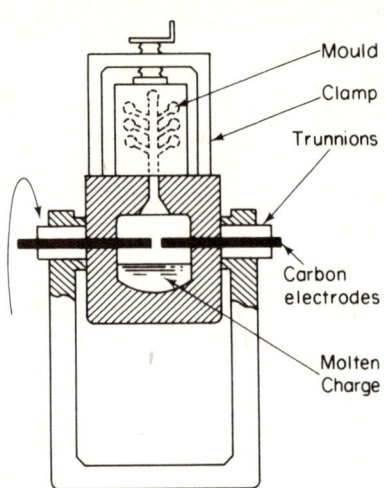

Figure 2.5 Tilting electric arc furnace

the mould, compressed air at light pressure may be applied, especially where fine and intricate detail is to be reproduced. After solidification the furnace is returned to its normal position and the mould released, followed by knocking out of the casting using chisels and pneumatic hammers.

Advantages of investment moulding

The advantages of the process may be summarised as follows:

(a) Castings have high dimensional accuracy with a tolerance of ± 0.08 mm possible.
(b) Good surface finish which may be such as to eliminate any further finishing such as machining. This is particularly significant where castings are to be produced in metals that are difficult to machine such as the nimonics (nickel, chromium alloys) as used for turbine blades.
(c) Since the wax pattern is an exact replica of the finished casting, including features such as holes and hollow cavities, etc., no cores or loose pieces are required.
(d) Components may be made as a single part which, if made by another casting process, might otherwise have to be made in sections and then assembled.

The main disadvantage of the process is that the tooling and operating cost is high. Since in many cases further machining will not be required, unlike castings produced by sand and shell moulding, this cost is often justified.

DIECASTING

The casting methods so far discussed use a refractory consumable mould, but the diecasting process requires the use of permanent metal moulds known as *dies*. Since these are expensive to design and manufacture and are used with high-cost machinery, diecasting can be an economic process only for high-volume production.

Diecasting metals

Metals used for diecasting are generally confined to a group of non-ferrous low melting point alloys, so that die life is kept to a maximum. Two important requirements of these alloys are that they possess good fluidity and do not suffer from 'hot shortness', a term used to describe brittleness of castings at elevated temperatures. The alloys used include those based on aluminium, zinc, magnesium, tin and lead, and, to a limited extent, brass and bronze.

By far the most common metals used are the aluminium- and zinc-based alloys. Typical composition for a diecasting aluminium alloy is 4% copper, 5% silicon, 3% iron, 2% nickel and 0.5% magnesium. Aluminium diecastings are used where a high strength-to-weight ratio is required. A typical zinc-based alloy consists of 4% aluminium, 2.7% copper and 0.03% magnesium. An alloy of this type has good casting properties and has the advantage that it requires a lower casting temperature compared with aluminium-based alloys. Furthermore, zinc is cheaper than aluminium. The use of lead and tin based alloys is rather limited, their main application being in the manufacture of low-pressure bearings and where strength is not an important factor. Magnesium alloys, sometimes known by the trade name as 'Elektron', are the lightest of all alloys and are used where weight and corrosion resistance are the most important considerations.

Diecasting processes Diecasting consists basically of two types of process, these being

(i) Gravity diecasting (ii) Pressure diecasting.

GRAVITY DIECASTING
This process is similar to sand casting with the exception that the mould is made from cast iron or from a special alloy steel. Runners and risers are required but since the mould material cannot be broken away, as in sand casting, the mould must be designed so that it can be separated to release the casting.

The simplest gravity die mould consists of two split dies, but due to the complexity of many components these have to be designed to consist of a number of removable sections and cores. However, the mould designer will try to keep these sections to a minimum, not only to reduce the cost of the mould but also to reduce the mould assembly time before the next pour. In addition to making allowance for shrinkage when the casting cools, the designer must also provide a suitable venting system so as to avoid porosity. Venting is usually achieved by providing shallow grooves along certain joint faces (about 0.5 mm deep), their position and number being dependent on the nature and complexity of the casting.

So that the casting does not solidify too quickly the die is pre-heated (approximately 200°C) before the first pour, after which the heat retained by the die due to the hot metal is sufficient for subsequent castings. Since metal moulds are used it is possible to produce castings to a better dimensional accuracy and improved surface finish than can be obtained by sand casting.

PRESSURE DIECASTING
With this method of diecasting the molten metal is injected into a closed metal die under considerable pressure. Compared with gravity die casting, pressure diecasting has the following

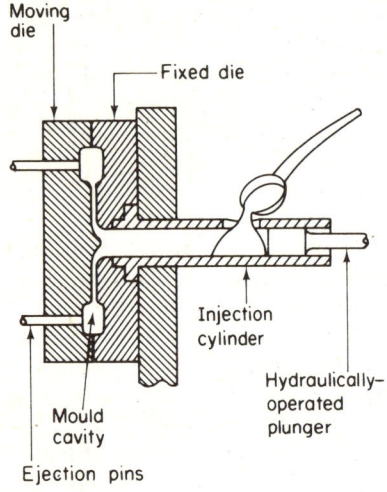

Figure 2.6 Cold chamber pressure diecasting

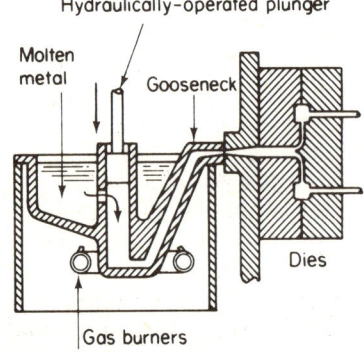

Figure 2.7 Hot chamber pressure diecasting

advantages:

(a) Thinner sections and finer detail may be produced.
(b) Better surface finish and improved dimensional accuracy.
(c) Improved grain structure due to metal being forced into die under pressure.

Pressure diecasting is grouped into two types of process, as follows:

(i) Cold chamber process (ii) Hot chamber process.

The method chosen depends mainly on the metal to be cast and the required production rate.

Cold chamber process: The essential features of the cold chamber machine are shown in Figure 2.6. In operation molten metal is transferred by hand-held ladle to the injection cylinder, after which it is forced into the die cavity by means of an hydraulically-operated plunger. The injection pressure applied varies according to the volume and type of metal to be injected but will be within the range of 14 to 70 MN/m^2.

After cooling, the dies are separated (or split) and the casting ejected, often assisted by ejector pins. The metals include, aluminium- and magnesium-based alloys, brass and bronze.

Hot chamber process: The hot chamber machine is illustrated in Figure 2.7. The machine consists basically of an injection cylinder immersed in a pot of molten metal, which is connected to the die by means of a gooseneck. In operation the plunger is lifted hydraulically so that a quantity of molten metal is permitted to enter the cylinder from where it is injected into the die cavity. The pressure used will be between 2.5 and 3.5 MN/m^2.

When the casting has solidified the dies are separated and the casting ejected. Opening, closing and injection are usually arranged in an automatic cycle, and with a fully-automatic machine it is possible to produce 2000 castings per hour. Aluminium-based alloys are seldom cast by the hot chamber process since this metal will react with the pot containing the molten metal (made from iron) to produce an 'iron pickup'. The process is therefore confined mainly to the casting of zinc alloys and magnesium alloys. Compared with the cold chamber process the cycle time is much faster.

PRESSURE DIECASTING DIES

The design and manufacture of dies call for specialist skill and a high standard of precision engineering and thus are very expensive to produce. This cost however is offset by the large number of castings that the dies will produce together with the short cycle time of the diecasting process.

The material from which dies are made must be extremely hard wearing so that maximum die life is obtained. For this reason special alloy die steels have been developed which contain a high percentage of chromium and vanadium as the main alloying elements.

In the design of a die, allowance must be made for shrinkage and ease of ejection. In additon to providing a slight draught to ease ejection, ejector pins are also provided, but these must be

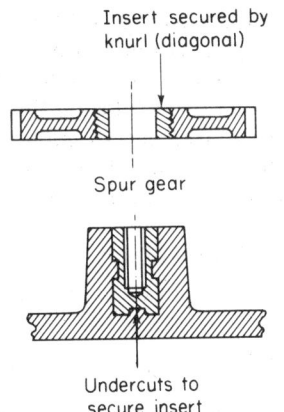

Figure 2.8 Diecast inserts

positioned with care so as not to damage or distort the casting when it is pushed from the die. Cooling is also essential and is achieved by allowing water to circulate through channels within the die.

Basically three types of dies are used, these being

(i) Single cavity (ii) Multi-cavity (iii) Combination.

Single cavity dies are the simplest type and are designed to produce one casting per cycle. *Multi-cavity dies* consist of a number of identical impressions and are used where the production rate is high. As many as 120 castings may be produced in one cycle using multi-cavity dies. *Combination dies* consist of a number of impressions, but not all of the same component, and are generally used where small parts for a complete assembly can be produced in one cycle.

Diecast inserts   Many diecastings require features such as tapped holes, splines and keyways to be incorporated into their design. Although these features can be machined into the casting it is often more practicable and economic to produce them in the form of an insert. It is essential for these inserts to be firmly attached to the surrounding cast metal, as shown by some examples in Figure 2.8. It is also important to ensure that inserts are located within the die in such a way that they cannot be displaced by the inrush of molten metal.

CHARACTERISTICS OF DIFFERENT CASTING PROCESSES

The characteristics of the casting processes that have been described may be summarised as in the following Table.

Casting process	Material for which suitable	Accuracy obtainable	Surface finish	Minimum section cast	Remarks
Sand casting	All metals	± 1.0 mm	Poor	3.0 mm	Consumable sand mould used. Due to high cost of mould preparation, process confined mainly to small quantity production.
Shell moulding	All metals	±0.10 mm	Improvement compared with sand casting	1.5 mm	Consumable mould used. Despite high cost of patterns, the short cycle time makes the process suitable for large quantities.
Investment moulding	All metals	±0.08 mm	High	0.8 mm	Consumable mould used. Equipment and operating costs high. No cores required. Suitable for high melting point metals.
Gravity diecasting	Most metals but mainly confined to aluminium and copper alloys	±0.08 mm	High but not as good as pressure diecasting	2.0 mm	Steel or cast iron split moulds used. Low operating cost but cost of dies relatively high.
Pressure diecasting	Low melting point alloys, e.g. alloys based on aluminium, zinc, tin, lead, magnesium	±0.04 mm	High	0.5 mm	Steel moulds used but due to their high cost, plus the cost of the operating equipment, the process is suitable only for large-volume production.

POWDER METALLURGY

Compared with shaping processes, such as casting and forging, the industrial use of metal powders for the manufacture of components is relatively new. Yet it is interesting to note that Russian coins were made from platinum powder during the early 1800s. In the past the components produced by this process included self-lubricating bearings, carbide tool tips, lamp filaments and magnets, but nowadays the method is used for a much wider range of applications.

Basically the process involves three distinct stages:

(a) Manufacturing of the powder
(b) Compacting (pressing) the powder
(c) Sintering the compacted component.

Manufacture of the powder

The method used to manufacture the powder depends on the metal involved and the particle shape and size required. Many techniques are used including the reduction of oxides and electrolysis. The most common technique is the atomisation of molten metal whereby an air jet is allowed to impinge on a stream of molten metal thus causing the metal to solidify rapidly into small particles. This method is suitable for a wide range of metals. After initial production, the powder grain sizes range from about 0.06 to 0.5 mm. To improve their pressing characteristics powders are mixed with additives such as a stearate of lithium or zinc.

Compacting the powder

Compacting or pressing the component to shape is carried out with a specially designed punch and die. The die is filled with a predetermined amount of powder and the punch stroke regulated to ensure that the component is compacted to the correct size and density. To ensure uniform pressure distribution within the compact a top and bottom punch are used simultaneously, the pressure used being within the range of 280 to 700 MN/m^2. The stages in compacting a plain bush are shown in Figure 2.9.

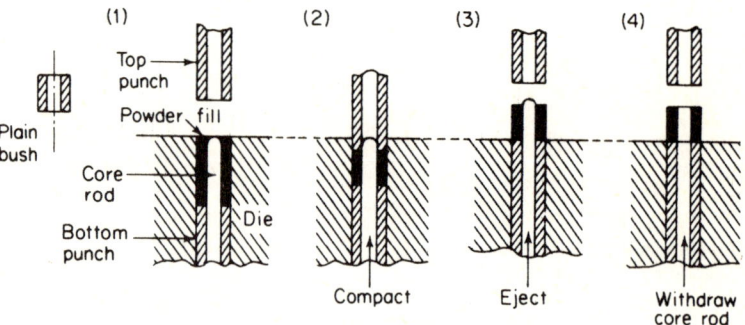

Figure 2.9 Stages in compacting a plain bush

Sintering

After pressing, the compacted component is relatively weak. To increase its strength the component is subjected to a heat treatment known as *sintering* whereby the powder particles become firmly attached to each other throughout the structure. The sintering temperature is dependent on the type of metal powder used but normally it will be just below the melting point of the metal, e.g. iron or steel compacts are sintered at about

1300 °C for a period up to one hour. A controlled-atmosphere furnace is essential so that oxidation is prevented, and in the case of iron/carbon alloys, decarburisation. The most common atmosphere used consists of hydrogen and nitrogen (cracked ammonia).

The advantages offered by powder metallurgy techniques may be summarised as follows:

(a) Closely controlled tolerances possible, e.g. ±0.02 mm.
(b) No finishing operations required such as machining.
(c) Possible to produce alloyed components which are otherwise difficult to manufacture using conventional melting techniques due to large temperature differences, e.g. tungsten (3400 °C) and copper (1083 °C).
(d) Possible to produce components having special properties, e.g. filters and porous (self-lubricating) bearings.
(e) Is suitable for producing components containing blind recesses with sharp corners which is otherwise difficult using conventional machining operations.

Limitations and design considerations

Parts produced by powder metallurgy techniques must be designed to suit the process, and since they are compacted vertically they must be of a shape to facilitate their ejection from the die. This means that re-entrant shapes or undercuts and cross holes cannot be produced unless machined after sintering. Furthermore, it is important that parts do not have any sharp or feathered edges. Examples of parts redesigned to suit powder metallurgy processes are shown in Figure 2.10.

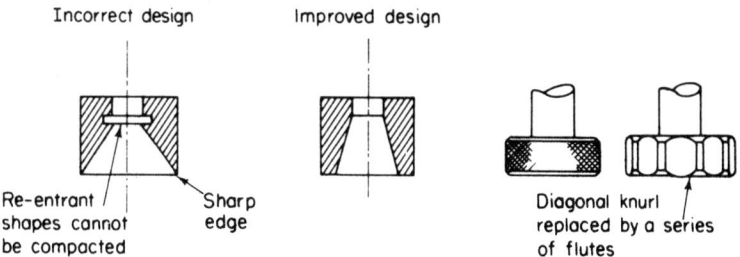

Figure 2.10 Design of sintered parts

SUMMARY

Moulds for shell moulding are made in two halves from a sand resin mix and are formed to shape by using heated metal half-patterns. After curing, the half-moulds are clamped together ready for pouring.

Investment moulding (lost wax process) requires a wax pattern of the component to be cast which is produced from a split die made from either plaster or metal. The wax pattern is covered with a refractory material (invested), and when this is cured the wax is melted out to reveal the mould cavity ready for pouring. No cores are required with this process.

Diecasting is used for producing light, intricate castings in low melting point alloys. The process is classified as being of the gravity or pressure type. Features such as splines, tapped holes and keyways are often included in die castings in the form of inserts.

Components produced from metal powders are first compacted (pressed) to shape then subjected to a heat treatment process known as sintering. Components produced by this process must not have any undercuts or re-entrant shapes, since these will make it impossible to eject the compacted component from the die.

QUESTIONS

(1) Using suitable sketches, describe the principle of shell moulding.

(2) Compared with sand casting, outline two advantages that shell moulding may offer.

(3) Using three separate headings, explain the stages in producing a casting by investment moulding.

(4) Using a suitable sketch, explain the principle of operation of the hot chamber pressure diecasting process. Why are aluminium alloys not suitable for this process?

(5) Diecasting processes are classified as being either

    (i) Gravity   (ii) Pressure processes.

Explain the difference between these two processes and state their main areas of application.

(6) Using suitable sketches, show two examples of the use of diecast inserts. In each case show clearly how the insert is secured to the surrounding metal.

(7) List three types of alloy used for diecasting and state two features that these alloys should possess if good cast impressions are to be produced.

(8) Outline the factors that the production engineer must consider before choosing a casting process.

(9) With reference to powder metallurgy, what is meant by the following terms:

    (i) Compacting   (ii) Sintering.

(10) Giving reasons why, name two components that would be suitable for manufacture from metal powders.

3 MEASUREMENT

INTRODUCTION In any form of manufacturing activity, measurement plays an essential part and will be required at all stages of manufacture, from the time the raw material is first processed to the final inspection of the finished product. Readers will no doubt be familiar with the more basic measuring techniques and equipment from their studies at level 2. The purpose of this chapter, therefore is to develop this knowledge further and deal with some of the more specialised applications of measurement commonly found in engineering production.

COMPARATORS Altough the dial test indicator is often used as a comparator there are specific instruments known as *comparators* which are based on different physical principles. The function of a comparator in precise measurement is to compare the size of the work with that of a standard of known size, such as a slip gauge or precision roller. The main feature of the comparator is that it employs some magnifying device which will give an accurate indication of the size difference between the standard and work being produced.

There are many different designs of comparator available, but basically they fall into one of the following groups:

(i) Mechanical, (ii) Electric, (iii) Optical, (iv) Pneumatic.

Mechanical comparators The two most common instruments in this group include

(i) The sigma comparator, (ii) The twisted strip comparator.

THE SIGMA COMPARATOR
The operating mechanism of this instrument is shown in Figure 3.1 (a).

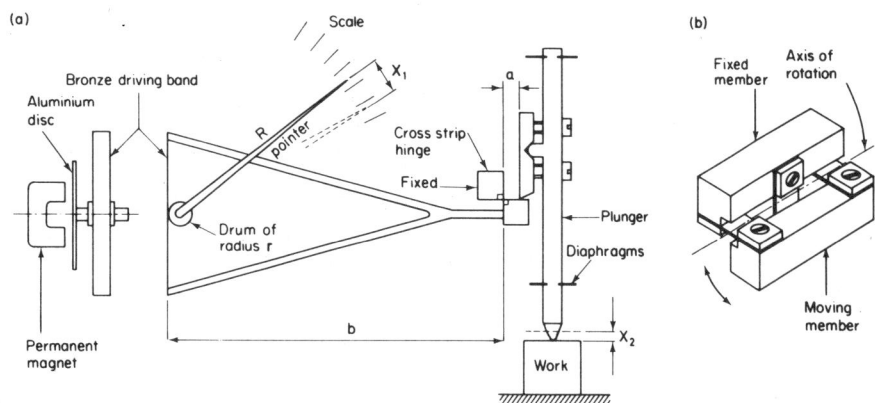

Figure 3.1 (a) Sigma mechanical comparator. (b) Details of cross-strip hinge

When the work or standard makes contact with the stylus, this causes the plunger to move vertically actuated by two metal diaphragms. This movement is transmitted to a cross strip hinge which converts it into a small angular rotation. The operating principle of the cross strip hinge will be readily understood by studying Figure 3.1(b). Since the Y arm is fixed to the moving member of this hinge its movement will be greatly magnified thereby producing an initial magnification of b/a. Attached to the end of the Y arm is a phosphor bronze driving band, wrapped around a drum and attached to the indicating pointer. Therefore, as the Y arm rotates the bronze band will drive the drum and hence rotate the pointer relative to the fixed scale. Due to the radius of the driving drum and pointer a further magnification will be produced by R/r. The overall magnification of the instrument will then be given by:

$$\frac{\text{Pointer movement}}{\text{Stylus movement}} = \frac{X_1}{X_2} = \frac{b}{a} \times \frac{R}{r}$$

In mechanical instruments of high magnification there is a tendency for the pointer to oscillate before coming to rest, which means that some form of damping device must be employed. With the sigma comparator this is achieved by eddy current damping, as shown in Figure 3.1(a). A thin aluminium disc is attached to the driving drum which rotates in close proximity to a permanent magnet. In so doing the magnetic field of the magnet will be cut, thereby inducing a small eddy current into the disc, around which will be produced another magnetic field. This field will act in opposition to the permanent magnet's field and hence will cause the pointer to come to rest quickly.

THE TWISTED STRIP COMPARATOR
The operating mechanism of this comparator is shown in Figure 3.2.

This comparator consists basically of a thin metal strip which has been given a series of twists along its length and onto which is fixed the pointer. If this twisted strip is put under tension, and therefore lengthened, the nett effect is that the pointer will rotate about the strip's axis. When the tension is released the pointer will return to its original position. The tension is applied to the strip by a strut and cranked lever device operated by the vertical movement of the central plunger. This plunger, which also carries the measuring stylus, operates through two thin metal diaphragms in a fashion similar to the sigma comparator.

The reader will no doubt have noticed in the twisted strip and sigma comparators the extensive use made of metal flexure devices, e.g. cross strip hinges and metal diaphragms, etc. These devices are frequently used in precision measuring mechanisms where a small, controlled movement is required since they offer the following advantages:

(a) No friction or wear.
(b) Easy to manufacture.
(c) Not dependent on high dimensional accuracy.

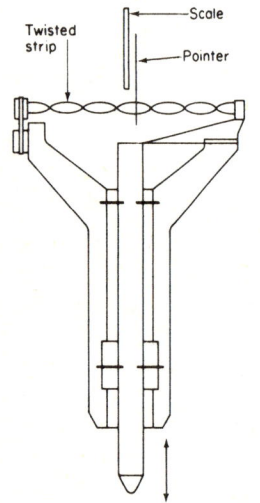

Figure 3.2 Twisted strip comparator

Electric comparators The principle of these comparators is to convert the linear displacement of the measuring stylus into an electric output.

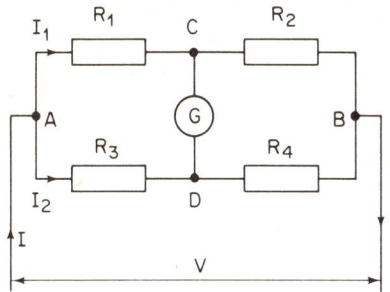

Figure 3.3 Wheatstone bridge network

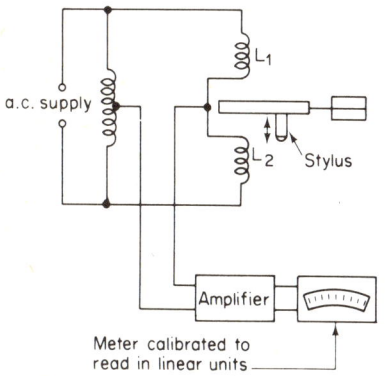

Figure 3.4 Electric comparator

Different types of circuit have been used for this comparator, but perhaps the most common type is that based on the Wheatstone bridge network. This circuit (Figure 3.3) consists basically of two series resistors, R_1 and R_2, connected in parallel with series resistors R_3 and R_4.

With reference to Figure 3.3, current I from the supply will divide at A, part flowing through R_1 and R_2 and part flowing through R_3 and R_4. These two currents I_1 and I_2 will meet at B to re-combine to form the return current I. Depending on the individual values of these resistors there will be somewhere between A and B two points where the drop in potential will be the same in both branches. These two points are represented by CD. If a galvanometer G is connected to these points then no current will be detected and under these conditions the circuit is said to be in balance whereby the following condition will exist

$$\frac{R_1}{R_2} = \frac{R_3}{R_4}$$

If one of these resistors is now varied then this balance will be disturbed and a current will flow through CD which will be detected by the galvanometer. By arranging the variable resistor to form part of the measuring head mechanism, the galvanometer may be calibrated to read in linear units.

The bridge circuit so far described is suitable only for d.c. supplies. In fact most electric comparators operate from a.c. supplies and although the operating principle is the same for direct current certain modifications to the circuit are required. The main difference being that resistors are replaced by inductors, two of which are incorporated in the measuring head. As the measuring stylus moves vertically, an iron armature will move between the inductors L_1 and L_2, thus causing the circuit to go out of balance. The change in current caused by this disturbance will be detected by a meter calibrated to read in linear units. A simplified circuit for use with alternating current is shown in Figure 3.4.

Electric comparators are extremely sensitive and are capable of high magnifications up to $\times 30\,000$.

Optical comparators

The operating principle of a typical instrument is illustrated in Figure 3.5.

In operation an illuminated datum line is projected onto a graduated screen via a small tilting mirror. When the measuring stylus moves vertically, the lever to which it is attached tilts this mirror and in so doing causes the reflected image of the datum to move relative to the screen. The magnification of the instrument may be considered in two stages.

1st stage: Magnification due to lever $= \dfrac{b}{a}$

2nd stage: Optical magnification $= \dfrac{2L}{c}$

Therefore, overall magnification $= \dfrac{b}{a} \times \dfrac{2L}{c}$.

The factor of 2 is brought about by the optical lever effect of the reflected ray, i.e. the reflected ray will move through twice the angle of tilt of the mirror.

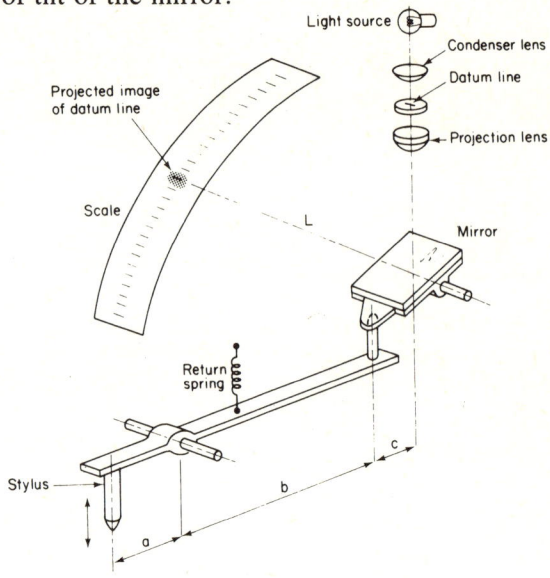

Figure 3.5 Optical comparator

Pneumatic methods of comparative measurement

The use of pneumatic principles for comparative measurement covers many applications and is generally regarded as being a versatile measuring technique, not only for the measurement of linear dimensions but also for geometric form. The principle of a typical system, sometimes known as a *back-pressure system*, is illustrated in Figure 3.6.

If air at constant pressure p_1 is permitted to enter a chamber then the pressure p_2 inside this chamber will depend on the relative sizes of the orifices A and B. Orifice A is known as the control jet and orifice B the measuring jet. If the part to be measured is brought into close proximity to orifice B, as indicated by the dimension x (Figure 3.6), then the pressure p_2 will increase. Therefore orifice B is behaving as if its cross-sectional area has been reduced, and the corresponding increase in pressure of p_2 can now be detected to represent the dimension x.

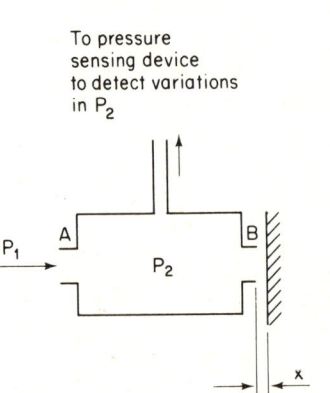

Figure 3.6 Principle of back-pressure type of pneumatic comparator

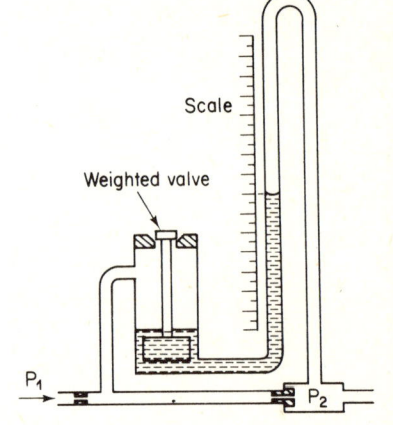

Figure 3.7 Air controller for pneumatic comparator (Solex Gauges Ltd)

An essential requirement of all pneumatic measuring systems is that air of constant pressure must be used. One method of achieving this is illustrated in Figure 3.7. Air from a factory air line or compressor is passed through a reducing valve and filter and enters a chamber containing a weighted valve partially immersed in a low-viscosity oil. The weight of this value determines the input pressure p_1 by allowing excess air to escape between the valve and its seating. The system incorporates a manometer tube so that variations in p_2 can be detected relative to a scale calibrated in linear units. The pressure used with this method is relatively low, in the order of about 8 N/m^2.

Methods of measurement The back-pressure system may be applied in one of three ways, as follows:

(i) By restriction (ii) By direct measurement
(iii) By relayed measurement.

Measurement by *restriction* is the simplest method and is used for measuring the diameter of small holes whereby the work effectively becomes the measuring jet.

With the method of *direct measurement* the measuring jet is of a fixed size and any variation in p_2 is caused by the gap between the jet and the work surface. If the work is close to the jet then the rate at which air escapes will be low, thus resulting in a high back pressure. This method finds many applications, as illustrated in Figure 3.8.

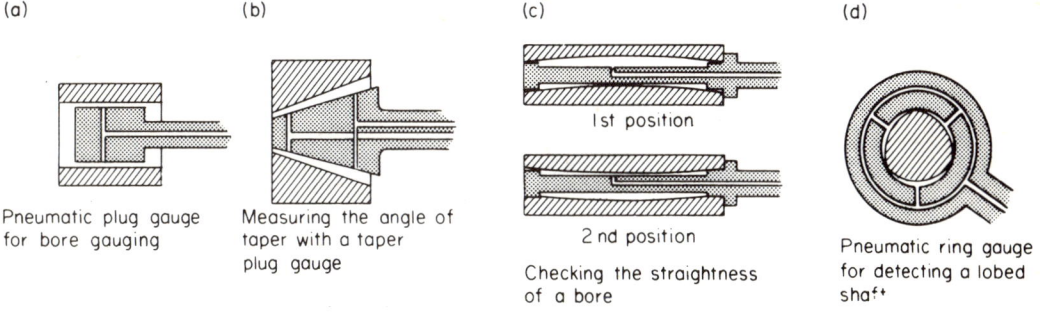

(a) Pneumatic plug gauge for bore gauging
(b) Measuring the angle of taper with a taper plug gauge
(c) 1st position / 2nd position — Checking the straightness of a bore
(d) Pneumatic ring gauge for detecting a lobed shaft

Figure 3.8 Applications of pneumatic gauging

The *relayed measurement* method is frequently used in the form of a comparator. Basically the measuring head consists of a valve which lifts off its seating due to the movement of the measuring stylus. The resultant back pressure will therefore be dependent on the gap between the valve and its seat. A typical pneumatic comparator together with a section through the measuring head is illustrated in Figure 3.9.

An alternative form of pneumatic measuring system, but one which requires a higher pressure (about 90 N/m^2), is known as the *flow-velocity type*. Air maintained at constant pressure by a regulator is admitted to a tapered glass tube. This tube contains a small, light float whose position depends on the rate of air flow within the tube, the rate of flow being controlled by the rate of escape of air from the control jet. The lower the flow rate the lower the float will be within the tube. The sensitivity

38 Measurement

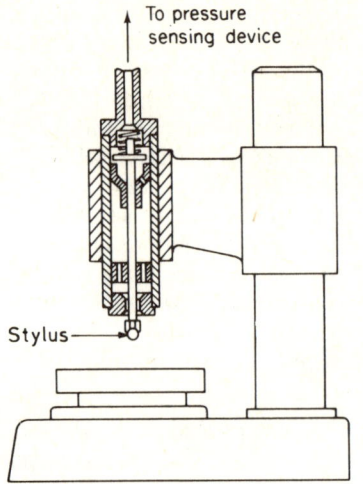

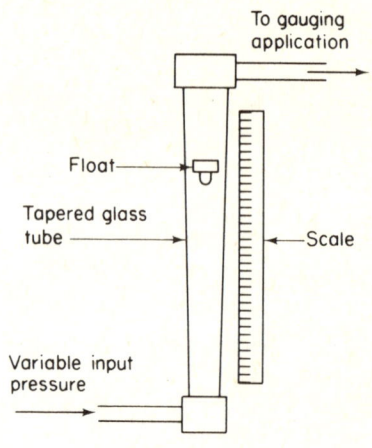

Figure 3.9 Pneumatic comparator

Figure 3.10 Flow-velocity pneumatic measuring system

of the instrument is governed by the degree of taper of the glass tube. The flow velocity instrument is illustrated in Figure 3.10.

The advantages of using pneumatic principles for comparative measurement may be summarised as follows:

(a) Wide range of features may be measured.
(b) Measuring head may be remote from the air controller, only a flexible tube being required to connect the two.
(c) Capable of high magnifications, × 20 000 being a typical value, while up to × 100 000 is possible.

OPTICAL PROJECTION   Optical projection is a convenient method of measuring both linear and angular dimensions and is ideally suitable for the measurement of complex forms which would otherwise be difficult to achieve with other methods. The principle of optical projection involves illuminating the work, whose image is then projected in the form of a shadow onto a suitable screen. The optical system in its simplest form is illustrated in Figure 3.11.

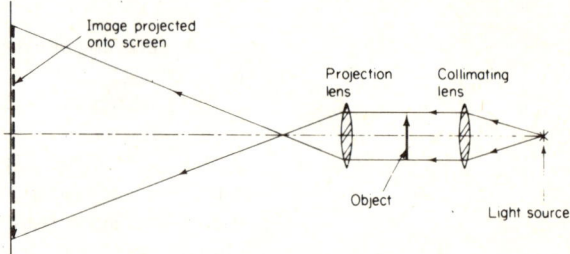

Figure 3.11 Simplified optical system of a projector

The collimating lens   The function of this lens is to concentrate the light source into a parallel beam of light. In measuring projectors it is important for the work to be illuminated by parallel (collimated) light in order to project the true dimension. By studying Figure 3.12 the reader will readily understand this principle.

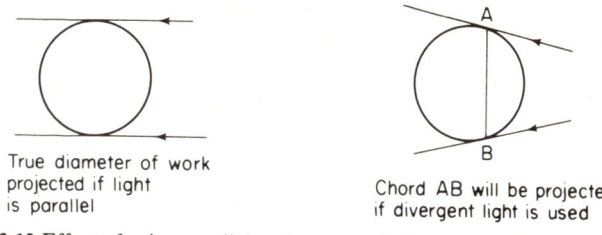

Figure 3.12 Effect of using parallel and non-parallel light

The projection lens

The function of this lens is to project an image of the work onto a suitable screen at an enlarged magnification. Typical magnifications used include × 10, 15, 25, 50 and 100. The projection lens shown in Figure 3.11 is a simple lens which suffers from a number of optical defects thus making this type of lens unsuitable for engineering projection applications. However, these defects may be reduced considerably by careful optical design which results in a projection lens system consisting of a number of elements (lenses), as illustrated in Figure 3.13.

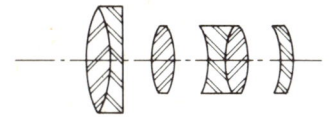

Figure 3.13 Construction of a typical projection lens

Types of projector

Early projectors consisted of a means of holding the work (workstage) and into which was incorporated the projection lens and the source of illumination. The screen was fixed to a wall onto which was projected an image of the work in a similar fashion to the home slide projector. This system has the

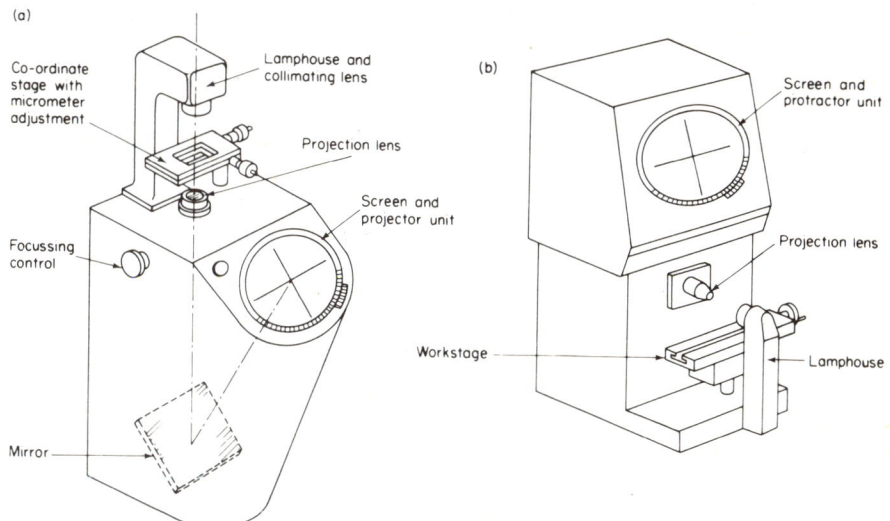

Figure 3.14 Cabinet projectors, (a) vertical type, (b) horizontal type

disadvantage that blackout conditions are essential and that a large floor area is taken up. Modern projectors however have a completely enclosed optical system which is 'folded up' and housed in a suitable cabinet. Cabinet projectors may be of the vertical or horizontal types, as shown in Figure 3.14(a) and (b) respectively.

Methods of measurement The simplest method of measuring a projected image is by using a steel rule. With reasonable care a steel rule can be used to measure to an accuracy of 0.3 mm. Although by precision measurement standards this accuracy is low, the true measured accuracy of the work will effectively be increased according to the magnification used. This means, for example, that when using a magnification of × 15 the real accuracy related to the work will be increased to $0.3/15 = 0.02$ mm.

For convenience and better reliability it is usual to measure linear dimensions with a co-ordinate stage, as shown in Figure 3.14(a). This device permits the work to be moved by a controlled amount in two directions at 90° to each other in the horizontal plane. In use the image of the work is first positioned against a datum in the form of a cross line on the screen and a reading on the relevant micrometer noted. The stage is then moved through the distance to be measured, relative to the datum, and a second micrometer reading noted. The difference in these two readings will therefore be the measured dimension. Angular dimensions may be dealt with in a similar fashion. This time a protractor screen unit is used which permits the screen to be rotated by a controlled amount using either a micrometer or Vernier adjustment, as shown in Figure 3.15.

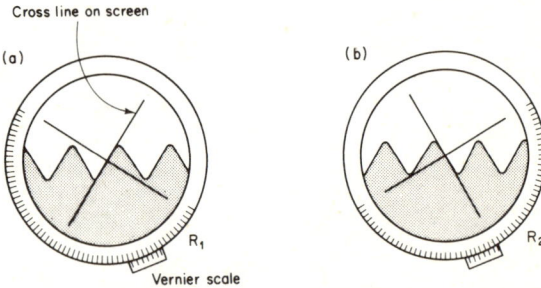

Figure 3.15 Protractor screen used to measure included angle of a thread. (a) Reading R_1 obtained by setting cross-line against one flank, (b) R_2 obtained by rotating protractor to line up cross-line against opposite flank. Thread angle = $R_1 - R_2$

The projection of complex profiles Optical projection is frequently used to inspect components of complex form, e.g. form tools and profile gauges. This is often achieved by comparing the projected profile with a template. This is specially-prepared by accurately drawing an enlarged profile (corresponding to the optical magnification) onto translucent drawing film, which is usually mounted between glass for protection. It is common practice when preparing these templates to indicate the component tolerance zone, thus making it possible to establish visually if the component is within the specified limits of size.

When projecting helical forms such as screw threads and helical gears, the parallel light produced by the collimating lens will cause interference of the projected image. This interference will be indicated by one side of the image being out of focus, as shown in the case of the flanks of a screw thread in Figure 3.16(a). This may be overcome by swivelling the lamphouse through the mean helix angle of the thread so that the parallel

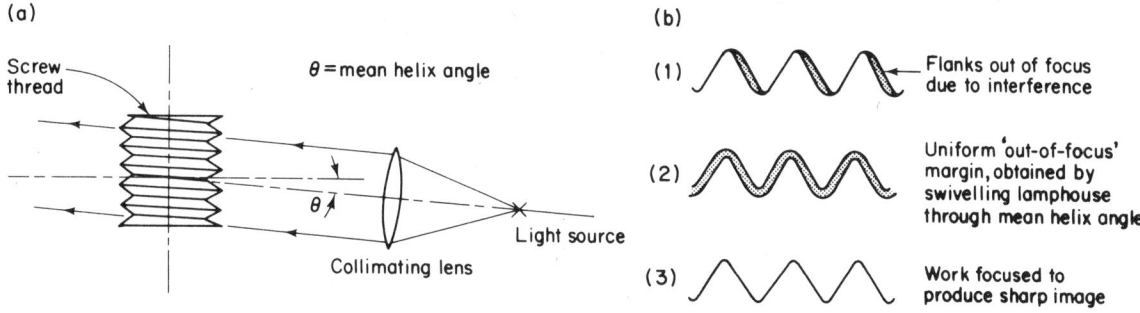

Figure 3.16 Projecting helical forms

light is effectively aligned to the thread helix. Although this lamphouse adjustment may be calculated, it is more usual to adopt the following procedure. First, deliberately put the image out of focus to produce an uneven out-of-focus margin around the thread. The lamphouse is then swivelled so that this margin becomes uniform, after which the image may be brought into focus to produce a sharp and well-defined profile. This procedure is illustrated in Figure 3.16(b).

Internal forms such as internal threads cannot be projected directly. The only way to overcome this problem is to make a cast in dental wax or plaster of Paris which can then be projected in exactly the same way as for external threads. In the case of internal threads it is important that such casts are lifted out and not unscrewed so as to minimise the risk of distortion. The procedure for making a plaster cast of the threads of a ring gauge is illustrated in Figure 3.17.

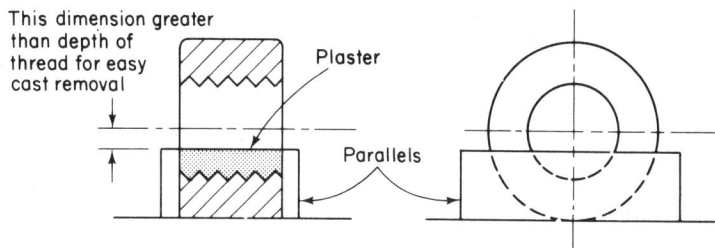

Figure 3.17 Making a plaster cast of the threads of a ring gauge

ANGLE MEASUREMENT The methods of angle measurement described in this section are confined mainly to the use of optical instruments. Used in conjunction with suitable accessories these instruments may be used for a wide range of applications.

The autocollimator The optical principle of this instrument is illustrated in Figure 3.18. If an illuminated source is positioned at the focal plane of a converging lens, then rays of light from this source will pass through the lens and be emitted on the other side as a parallel beam (Figure 3.18(a)). If a plane reflector is now positioned at 90° to this beam then the light rays will be reflected back along their previous path. However, if the reflector is given a slight

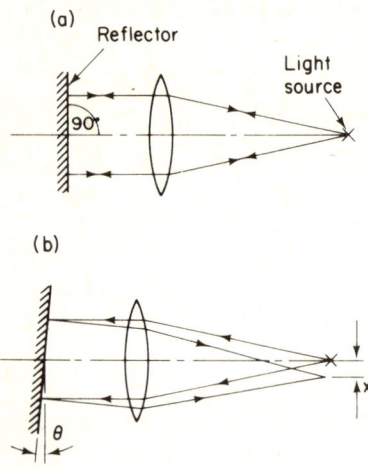

Figure 3.18 Principle of the autocollimator

angular tilt (θ) then the reflected rays will form an image at the focal plane, but displaced by the distance x (Figure 3.18(b)) from the optical axis. This linear displacement may be calibrated to represent the angular displacement of the reflector.

A practical version of the autocollimator is illustrated in Figure 3.19(a). The illuminated source takes the form of a target consisting of a cross line. Rays from this target are reflected from the beam splitter (a cube that partly reflects and partly transmits light) and pass through the object lens to the reflector. On reflection the rays then pass through the beam splitter where an image of the target is formed in the focal plane of the objective lens. In this plane, and controlled by the micrometer eyepiece, a pair of setting lines is used to measure the image displacement. The micrometer drum is calibrated to read directly in seconds of arc, of which one revolution represents 30 seconds. The range of the instrument is extended up to 10 minutes of arc by a micrometer turns-counter scale positioned over the eyepiece, one division on this scale representing one complete revolution of the micrometer drum. The use of the micrometer eyepiece is illustrated in Figure 3.19(b).

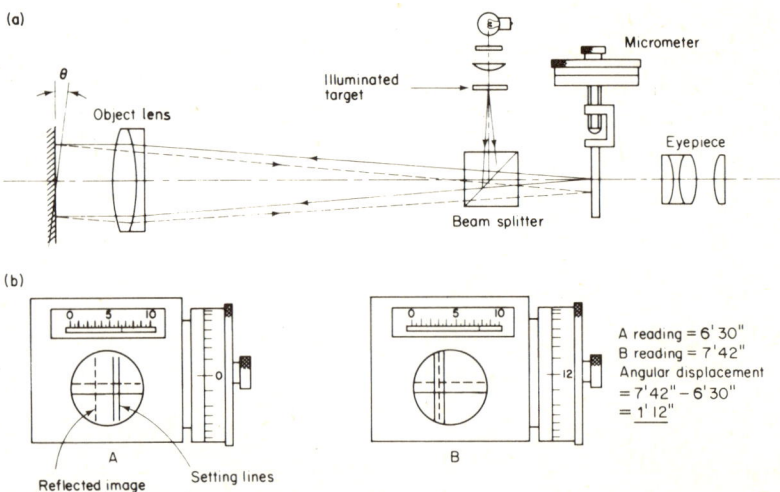

Figure 3.19 The autocollimator (practical version) (Rank Taylor Hobson)

Applications of the autocollimator Although primarily an instrument for measuring small angular diplacements, the autocollimator used in conjunction with suitable equipment is often used for the measurement of small

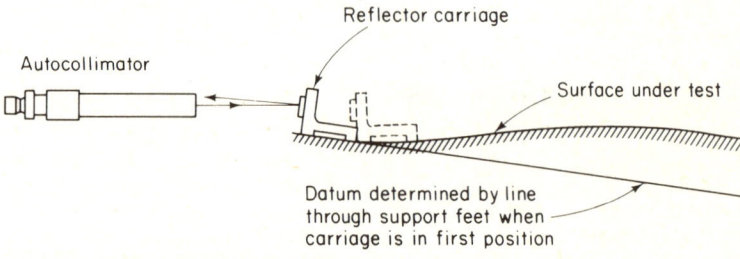

Figure 3.20 Straightness testing using an autocollimator

linear errors. One of the most important applications is in the measurement of straightness on structures such as machine tool guideways and large-surface tables. During a straightness test, readings are taken from a reflector carriage each time it is moved a distance equal to the pitch of its support feet. In this way any errors in straightness that are present will cause the reflector to tilt relative to the line of sight of the autocollimator. The set-up for the test is shown in Figure 3.20.

After a suitable number of readings have been obtained, the error in straightness is determined by using the following tabulated method:

1 Position	2 Autocollimator reading		3 Difference from first reading	4 Linear rise/fall	5 Cumulative rise/fall	6 Adjustment for zero	7 Errors from straight line
	min	sec	sec	(μm)	(μm)	(μm)	(μm)
			0	0	0	0	0
1	6	12	0	0	0	+0.2	+0.2
2	6	14	+2	+1.0	+1.0	+0.4	+1.4
3	6	15	+3	+1.5	+2.5	+0.6	+3.1
4	6	13	+1	+0.5	+3.0	+0.8	+3.8
5	6	11	−1	−0.5	+2.5	+1.0	+3.5
6	6	9	−3	−1.5	+1.0	+1.2	+2.2
7	6	8	−4	−2.0	−1.0	+1.4	+0.4
8	6	9	−3	−1.5	−2.5	+1.6	−0.9
9	6	11	−1	−0.5	−3.0	+1.8	−1.2
10	6	14	+2	+1.0	−2.0	+2.0	0

Columns 1 and 2: Indicated here are the positions of the reflector and autocollimator readings respectively.

Column 3: Since the first reading is an arbitrary value it is necessary to produce a zero datum and this is brought about by subtracting the first reading from all the other readings.

Column 4: The angular displacements in Column 3 must now be converted into linear displacements of the back support feet of the reflector, which may be a rise or fall depending on the direction of tilt. To do this, the amount of rise or fall for a tilt of 1 second of arc over the length of the reflector must be known. For the purpose of this example it is assumed that the reflector is 100 mm long, which means for an angular tilt of 1 second of arc the back feet will rise or fall 0.5 μm (0.0005 mm). Therefore the values in Column 3 are simply multiplied by 0.5 μm.

Column 5: The values here are cumulative values arrived at by algebraically adding each value in turn to the succeeding ones. This must be done since the datum exists as a straight line passing through both front and back feet of the reflector in the first position. This will be readily understood by refering to Figure 3.20. If these values are plotted into a graph then the outline of the surface will be shown as in Figure 3.21(a).

Column 6: Although the graph shape of the cumulative values represents the true outline, it is usual to draw this so that 'both ends' are at zero relative to the horizontal axis. This means an

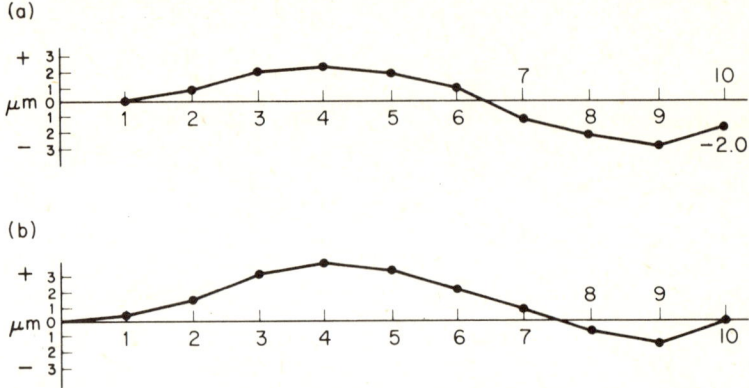

Figure 3.21 Results of straightness test in the form of a graph

adjustment must be applied to the values in Column 5, the basis of which is obtained by dividing the last value in Column 5 by the number of positions, i.e. $2/10 = 0.2\,\mu m$. Each adjustment will now appear in Column 6 in multiples of $0.2\,\mu m$ throughout the range. Since the last value in Column 5 is negative, it means that all the adjustment values must be positive so that the last reading in Column 7 becomes zero.

Column 7: Here each value in Column 6 is added to its respective value in Column 5 to obtain the true errors from a straight line, as shown graphically in Figure 3.21(b).

A further application of the autocollimator is in the measurement of squareness. A typical example appears in Figure 3.22(a) where it is required to check the squareness of the axis of bore A with the axis of bore B. To do this, tooling plugs are first inserted in these bores, their ends being made reflective by attaching a small flat mirror to each, usually held on with wax. The reader will no doubt appreciate that the ends of the plugs must be square with their axes and that the mirror must not only be flat but also have parallel faces. With the

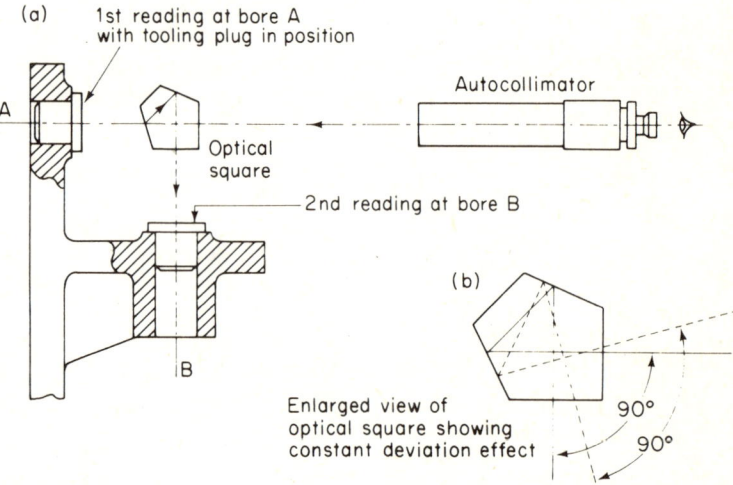

Figure 3.22 Squareness testing using an autocollimator

autocollimator in position a reading is obtained from bore A. An optical square, consisting of a pentagonal prism is now positioned over bore B, so that the line of sight of the autocollimator is turned through 90°. If the two bores are square with each other then a repeat reading will be obtained when looking at the mirror over bore B. The orientation of the prism is not critical, since this type of prism has the property of deviating the line of sight through 90° irrespective of the angle that the autocollimator axis makes with the front face of the prism. This effect is simply illustrated in Figure 3.22(b).

The angle dekkor The principle of this instrument (Figure 3.23) is identical to that of the autocollimator. It is mounted on its own small surface plate by an adjustable bracket. The measuring accuracy is not as high as that of the autocollimator, the direct reading accuracy being 1 minute of arc, while by careful estimation 0.2 minute of arc is possible. The illuminated target used in the autocollimator is effectively replaced by an illuminated scale on a glass screen

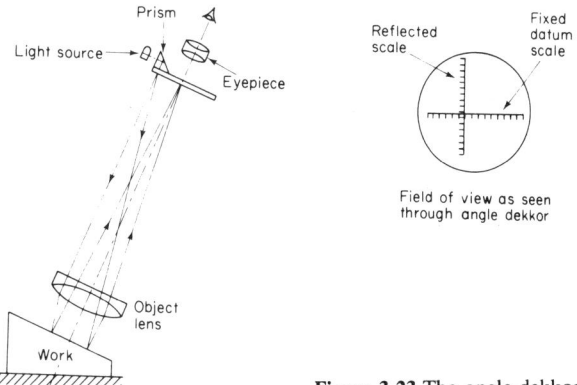

Figure 3.23 The angle dekkor

set in the focal plane of the object lens. The image of this scale is seen within the field of view of the eyepiece after reflection from the work surface. This image will be disposed relative to a horizontal fixed scale which serves as a datum. In use the instrument is used as a comparator, i.e. it must first be set to a known standard, usually in the form of angle gauges.

The following example will illustrate how the angle dekkor may be used to measure the included angle of a V block, shown in Figure 3.24, that has been machined nominally to 90°.

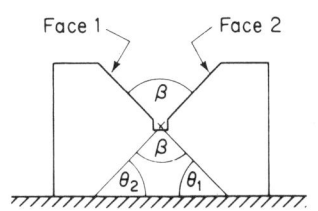

Figure 3.24 Measurement of the included angle of a V block

Example: Readings are taken from each face of the V after first setting the angle dekkor to angle gauges built up to 45°.
Datum reading from 45° angle gauge = 24′.
With angle gauge substituted for work:

Reading over face 1 = 28′
Reading over face 2 = 22′

Therefore, $\theta_1 = 45° \, 4'$, since the reading over face 1 compared with the datum reading is 'greater' by 4′.

$\theta_2 = 44° 58'$, since the reading over face 2 compared with the datum reading is 'less' by 2'.

Therefore, the included angle of the V (β) may be expressed as

$$\beta = 180 - (\theta_1 + \theta_2)$$
$$= 180 - (45° 4' + 44° 58')$$
$$= 180 - 90° 2'$$
$$\beta = 89° 58'$$

During the above procedure the necessary reflective surfaces, when taking readings with the V block, may be obtained by using a slip gauge or flat mirror.

The precision level

Although this is an instrument for measuring small angular displacements, the *precision level* is frequently used for straightness and squareness testing, especially on structures such as machine tools. The instrument basically consists of a ground glass tube, known as a *vial* that has been bent or ground to a large radius and partially filled with a suitable liquid so that an air bubble is formed. The principle of the instrument is based on the fact that this bubble will always remain at the highest point of the tube irrespective of its angular tilt. The sensitivity of the level is governed by the radius of this tube and is usually graduated so that each tube division represents an angular tilt of the base by 10 seconds of arc.

Precision levels are of two types, being either a *block level* or *square block level*, as illustrated in Figure 3.25.

The *square block level* is particularly useful for testing vertical surfaces as well as those lying in the horizontal plane. The *block level* is often used as an alternative to the autocollimator for the straightness testing of machine tool guideways.

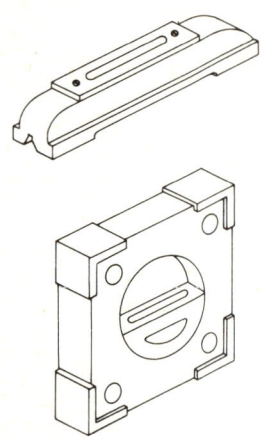

Figure 3.25 Precision block levels

SURFACE FINISH MEASUREMENT

Before the Second World War the assessment of surface finish was largely dependent upon personal skill, based on feel and visual inspection. While for some classes of work this is sufficient there are many cases where a more reliable method is required in the form of a numerical assessment. The measurement of surface finish has now become a well-established technique, due mainly to the need for improving fits between assembled parts, together with the demand for components to have high resistance to fatigue and corrosion.

No surface can ever be completely smooth and will always consist of small irregularities in the form of peaks and valleys. The texture produced by these irregularities is complex, but may however be classified into the following features:

Roughness (*primary texture*): This refers to the texture arising from the inherent characteristics of the finishing process and is associated with small scratches and toolmarks on the surface. Roughness is usually determined by the cutting speed and the quality of the cutting tool or abrasive used.

Waviness (*secondary texture*): This refers to the texture arising from characteristics not directly associated with the metal cutting process such as machine tool vibrations and tool chatter. On full examination of a machined surface it will be found that the surface roughness will be superimposed on the waviness, as shown in Figure 3.26.

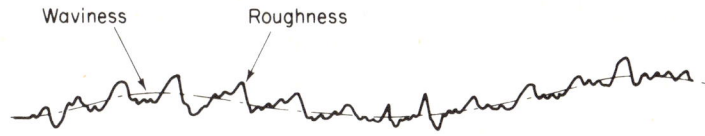

Figure 3.26 Surface finish characteristics

Lay This term is used to describe the direction of the surface pattern left by the finishing process. For example the characteristic pattern left by a shaping or planing operation will consist of a series of parallel feed marks, whilst for an endmilled surface the lay will be in the form of a series of circular arcs.

System of measurement In the past many systems for surface finish measurement have been adopted, but the one most universally accepted is known as the *arithmetical mean deviation* (R_a). This system simply expresses the numerical value of the surface as an average height of the surface irregularities, in micrometers (μm) from a mean centre line. When a magnified graphical recording of the surface is obtained the R_a value may be obtained by using the following method. By considering the recording shown in Figure 3.27 a centre line is first drawn through the trace to a suitable length, known as the 'sampling length'. This should be done in such a way that the sum of the areas above this line equals the sum of the areas below it. Furthermore, it must be appreciated that the sampling length should take into account those irregularities which are representative of the surface as a whole. The areas may be determined by using one of the mathematical rules, e.g. mid-ordinate rule, or by using a planimeter. If the areas are in square millimeters then the value for R_a may be calculated from the following expression:

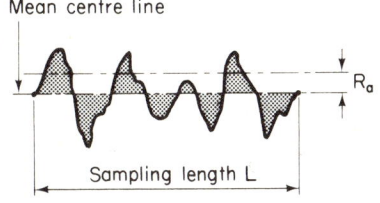

Figure 3.27 Arithmetical mean deviation (R_a)

$$R_a (\mu m) = \frac{\text{Sum of areas above and below centre}}{L} \times \frac{10^3}{V}$$

Where V = Vertical magnification
L = Sampling length

Methods of measurement The simplest method, although strictly speaking not a measurement, is the use of *standard comparative finishes*. These consist of specially prepared surfaces that have been calibrated to a known R_a value. They are available in sets representing different machining processes, each set consisting of about six surfaces from very fine to rough. They can be used by either making a visual comparison with the work or comparing by feel, usually with a fingernail. Although each standard finish has a known R_a value a certain amount of skill is required if an accurate comparison is to be made. The main advantage of this method is that these finishes are relatively cheap and ideally suited to workshop use.

Instruments employing a stylus These types of instrument incorporate a pointed stylus which is moved over the surface under test, and as it does so it will be displaced vertically by the surface irregularities of the test surface. Most instruments of this type employ a *skid* which usually consists of a curved shoe attached to the main body of the instrument and provides a datum for the stylus movement.

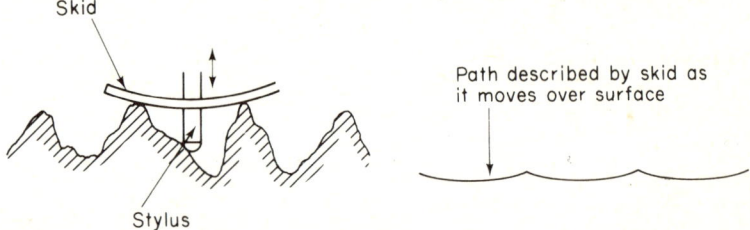

Figure 3.28 Use of the skid to provide horizontal datum for stylus displacement

The principle of the skid is illustrated in Figure 3.28. The vertical displacement of the stylus is magnified by using mechanical or electric methods.

Tomlinson recorder Although this mechanical instrument (Figure 3.29) has now been superseded by modern electric instruments, it is of interest since it enables a numerical value of surface finish to be obtained from first principles.

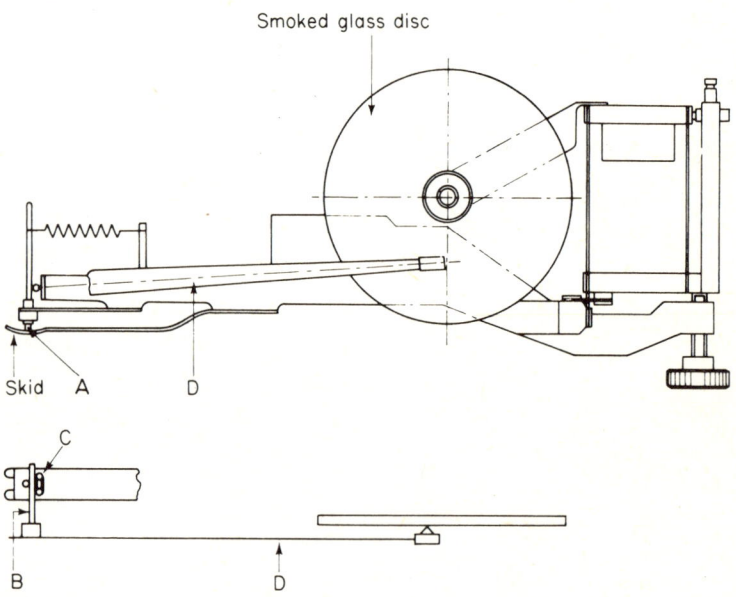

Figure 3.29 The Tomlinson recorder (J. E. Baty & Co Ltd)

As the stylus A moves vertically it will cause the spindle B to roll between the two fixed rollers C and the stylus. The spindle carries an arm D which has attached to its other end a diamond-tipped pointer which makes contact with a smoked glass disc. Therefore, as the stylus is drawn over the surface the pointer will respond to its movement by producing a profile of the surface on the smoked glass. The vertical magnification of the profile will be ×100, and since the horizontal movement of the stylus is not magnified, the profile will be compressed, and therefore must not be regarded as the true shape of the surface tested. Since the profile on the disc is very small it must be further magnified by optical projection from which a permanent record may be obtained. This is achieved either by exposing the projected image to photographic paper and developing as for a normal print, or simply by transferring the image onto tracing paper. The numerical value (R_a) for the surface may now be determined as outlined earlier in this section.

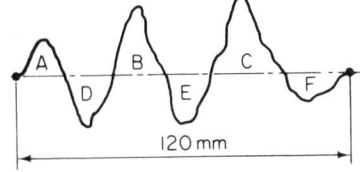

Figure 3.30 Profile trace from Tomlinson recorder

Example: The profile illustrated in Figure 3.30 shows a trace obtained from the Tomlinson recorder.

Given the following information determine the arithmetical mean deviation (R_a) for this surface.

Area (ref)	A	B	C	D	E	F
Area (mm²)	105	182	246	160	177	110

Optical magnification 50
Mechanical magnification 100
Sampling length (magnified) 120 mm

Total vertical magnification (V) of trace = 100 × 50
= 5000

$$R_a = \frac{\text{Sum of areas above and below centre line}}{L} \times \frac{10^3}{V} \mu m$$

$$= \frac{105 + 182 + 246 + 160 + 177 + 110}{120} \times \frac{10^3}{5000}$$

$$= \frac{980}{120} \times \frac{1}{5}$$

$R_a = 1.63$ Say, **1.6 μm**

Electric recording instruments

Electric instruments represent by far the most common method of measurement on account of their reliability, accuracy and ease of operation. The layout of a typical instrument is illustrated in Figure 3.31. A high-frequency alternating current having a constant waveform (known as a *carrier wave*) is applied to a bridge circuit containing two variables inductances. As the stylus is displaced vertically these inductances will change due to the variation in air gap between the pivoted stylus and the iron core. This in turn will cause the carrier wave to change (modulate) in response to the surface irregularities. The modulated signal is now amplified and demodulated so that the change in current is representative of the surface under test. The horizontal distance moved by the stylus on electric instruments is known as the *measuring traversing length* and contains several sampling lengths. In this way the stylus will explore most of the irregularities that are characteristic of the surface as a whole. The sampling length chosen will depend on

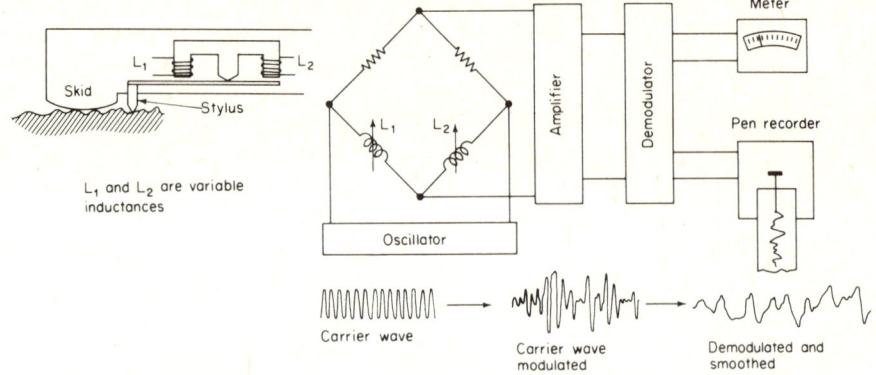

Figure 3.31 Electric surface-finish recording instrument (Rank Taylor Hobson)

the roughness of the surface for which the appropriate length will be chosen from one of the following values:

0.08   0.25   0.8   2.5   8.0   25 mm

ALIGNMENT TESTING OF MACHINE TOOLS

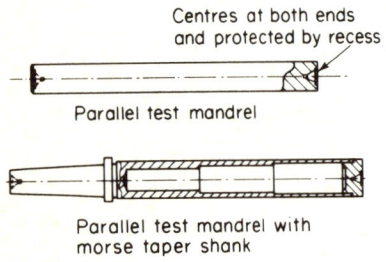

Figure 3.32 Test mandrels

In general, the use of machine tools makes a major contribution to the output of manufactured products. So that these machines can be relied upon to produce work of consistent quality it is necessary to perform certain tests on them to check the geometrical relationships between the various moving elements. Although these tests are carried out before the machine leaves the manufacturer, it will be necessary to repeat some of these tests before the machine is used since installation may cause distortion of the machine structure. Furthermore, after prolonged use, or after an old machine has been re-conditioned, it will again be necessary to carry out such tests. The tests designed for individual machines are numerous and in the space available in this volume only some of the more important ones can be considered.

The equipment used is often relatively simple, the most common being test mandrels and dial indicators used in conjunction with suitable accessories. *Mandrels* are used as a reference from which measurements can be taken and may be either of the parallel type for use between centres or parallel with a morse taper shank. Both mandrels are illustrated in Figure 3.32.

TESTS FOR CENTRE LATHES

Straightness of slideways

This test, illustrated in Figure 3.33, is the first test that should be carried out on a newly-installed lathe. Here a *precision spirit level* is most commonly used whereby measurements are made along the length of the slideway in different positions. Both front and rear slideways must be treated in this manner.

It is important that both slideways lie in the same plane and this is checked by positioning the spirit level transversely and then taking several readings along the length of slideway. When

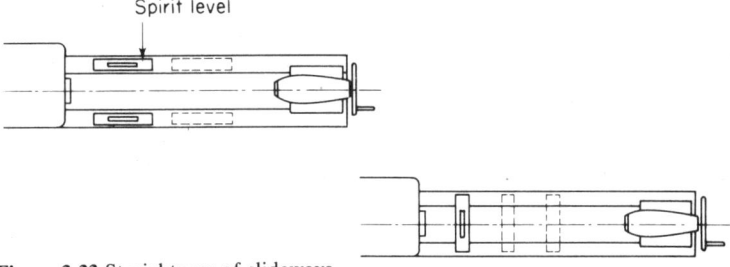

Figure 3.33 Straightness of slideways

the front and rear slideways do not lie in the same plane, i.e. the spirit level has revealed a twist; this is often referred to as 'crosswind'.

Spindle axis parallel with carriage movement

This test is illustrated in Figure 3.34. A parallel test mandrel with a morse taper shank is located in the spindle nose. A dial indicator is now mounted onto a convenient part of the carriage and adjusted to make contact with the mandrel. The carriage is then traversed along the length of the mandrel to reveal any out-of-parallelism. This test is carried out in both the vertical and horizontal planes.

Cross-slide movement square with spindle axis

The set up for this test is shown in Figure 3.35. A straightedge is clamped to the cross-slide and adjusted so that it is at 90° to the spindle axis. This is achieved by obtaining the same readings on a dial indicator (held in a chuck) when it is turned over from the front to the rear position. The straightedge therefore provides a 90° datum. With the dial indicator in contact with the straightedge, the cross-slide is then moved and any deflection on the dial indicator will reveal the cross-slide not moving square on the spindle axis.

Tailstock quill axis parallel with bed

This test is illustrated in Figure 3.36. A test mandrel with a morse taper shank is fitted into the tailstock quill. A dial indicator clamped to the carriage is now traversed along the length of the mandrel. This test is carried out in both the vertical and horizontal planes.

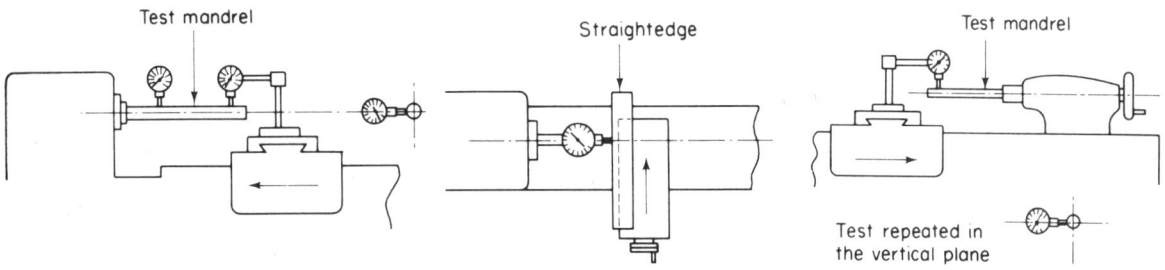

Figure 3.34 Spindle axis parallel with carriage movement

Figure 3.35 Cross-slide movement square with spindle axis

Figure 3.36 Tailstock quill axis parallel with bed

TESTS FOR HORIZONTAL MILLING MACHINES

Table movement parallel with centre T-slot

This test is illustrated in Figure 3.37. A dial indicator of the lever type is clamped to a convenient part of the machine and the stylus brought into contact on the inside vertical face of the slot. The table is then traversed along its entire length and any deflection on the dial indicator observed. On old machines the inside of this slot may have surface defects through continual use. Since these defects will be revealed by the dial indicator, they should however be disregarded during the test.

Spindle axis square with centre T-slot

This test is illustrated in Figure 3.38. A dial indicator of the lever type is attached to the machine spindle via an extension arm. The stylus of the dial indicator is brought into contact with

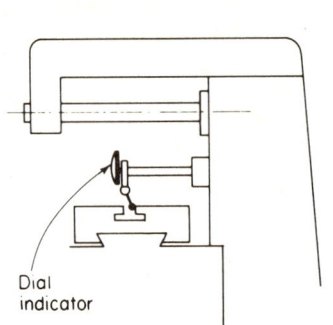
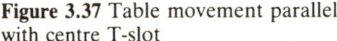
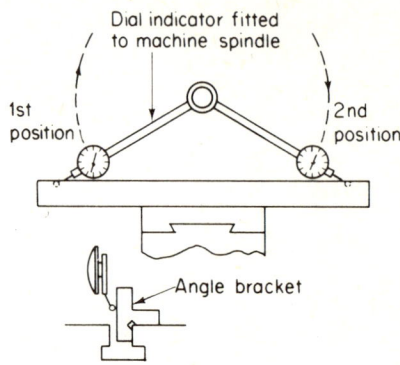

Figure 3.37 Table movement parallel with centre T-slot

Figure 3.38 Spindle axis square with centre T-slot

the vertical face of the slot at one end of the table. The spindle carrying the dial indicator is rotated so that the stylus makes contact with the vertical face of the slot at the opposite end of the table. If the spindle axis is square with the slot then the readings will be the same at both ends of the table. Due to the possibility of surface defects existing on the vertical face of the slot, it is usual to take the measurements relative to a ground angle bracket located against the side of the slot.

Squareness of table with vertical ways

There are two methods of performing this test, the choice depending on the accuracy required.

1st method (Figure 3.39(a)): A precision square is clamped to the table with its blade vertical, and a dial indicator (clamped to the vertical ways) adjusted to make contact with the edge of the square. The knee of the machine is then moved vertically, while observing any deflections on the dial indicator. The test should be repeated with the square at 90° to the first position to detect any out-of-squareness with the sides of the vertical ways.

2nd method (Figure 3.39(b)): On precision milling machines such as those found in toolrooms it may be necessary to carry out an optical squareness test. This involves obtaining a reading from an autocollimator when sighted onto a reflector carriage positioned on the machine table. A second reading is now obtained with the reflector against the vertical ways, the line of sight of the autocollimator being deviated through 90° by means of an optical square. If the vertical ways and table are square to each other then the two readings will be the same.

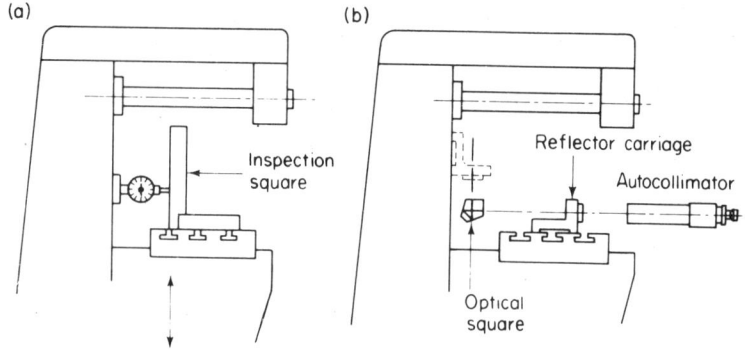

Figure 3.39 Squareness of table with vertical ways

TESTS FOR DRILLING MACHINES

Spindle axis square with table This test, illustrated in Figure 3.40, is one of the most important for drilling machines. A dial indicator fitted to a cranked arm and held in the chuck is adjusted to make contact with the machine table. The machine spindle is slowly rotated by hand and any deflection on the dial indicator being observed as it sweeps though a circle over the table.

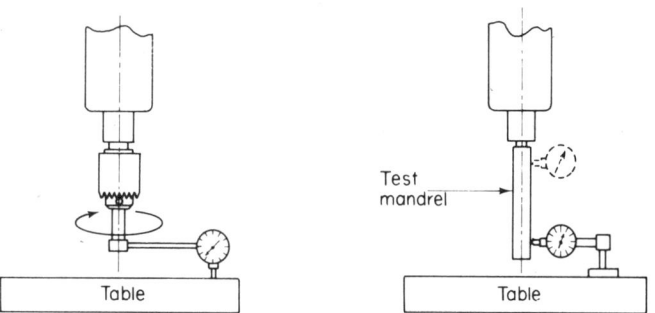

Figure 3.40 Spindle axis square with table   **Figure 3.41** Spindle axis runs true

Spindle axis runs true This test is illustrated in Figure 3.41. A parallel test mandrel with a morse taper shank is located into the machine spindle. With a dial indicator in contact with the mandrel, the spindle is slowly rotated by hand and any deflection on the dial indicator observed. This test should be carried out in the two positions shown.

LIMITS AND FITS Components manufactured to the same specification cannot all be made to the same size, since factors such as tool wear and operator fatigue will cause a slight variation in size. This means that at the design stage a permissible variation in size must be specified, but this should be such as will not impair either the quality or the function of the finished product. To meet this requirement a system of limits and fits is used which is designed to enable parts to be made economically and of uniform quality.

Furthermore, by adopting a system of limits of fits interchangeability becomes possible, thus dispensing with the need for selective assembly.

The essential features of a limits-and-fits system are illustrated in Figure 3.42.

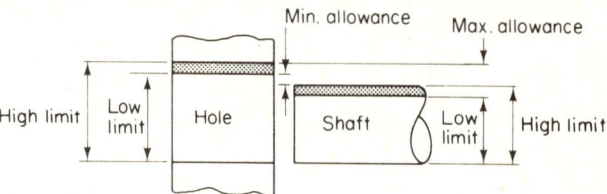

Figure 3.42 Limits and fits

The following definitions are used with reference to limits and fits:

Basic size: This is sometimes known as the 'nominal size' and it is the size of the part as laid down by the designer which is permitted to vary slightly.

Limits of size: Since no part can be manufactured exactly to the basic size, the designer must specify a permissible variation in terms of limits of size. When a part is manufactured to the smallest size this is known as the 'low limit' and conversely when a part is manufactured to the largest size this is known as the 'high limit'.

Tolerance: This is the difference between the high and low limits of size. Tolerances may be specified by using either a *unilateral* or a *bilateral system*. In the unilateral system the tolerance is disposed in one direction relative to the basic size.

For example: $82.00 \,{}^{+0.04}_{+0.01}$ mm and $82.00 \,{}^{-0.05}_{-0.02}$ mm

In the bilateral system the tolerance is disposed on either side of the basic size.

For example: $82.00 \,{}^{+0.02}_{-0.02}$ mm alternatively 82.00 ± 0.02 mm

Maximum and minimum metal condition: The *maximum metal condition* for a hole is when it is made to its low limit and likewise when the hole is made to its high limit it will be on its *minimum metal condition*. Conversely, for a shaft the maximum metal condition will be when it is made to its high limit and the minimum metal condition will be when it is on its low limit.

Allowance: This is the designed difference in size between two assembled parts. The allowance will be a maximum when both hole and shaft are made to their minimum metal condition. The allowance will therefore be a minimum when the hole and shaft are made to their maximum metal condition.

CLASSIFICATION OF FITS

Three types of fit are possible (Figure 3.43), these being known as

(i) Clearance (ii) Interference (iii) Transition.

A *clearance fit* will always be produced irrespective of the size (within the specified limits) to which the two parts have been made. The allowance on this type of fit will always be positive.

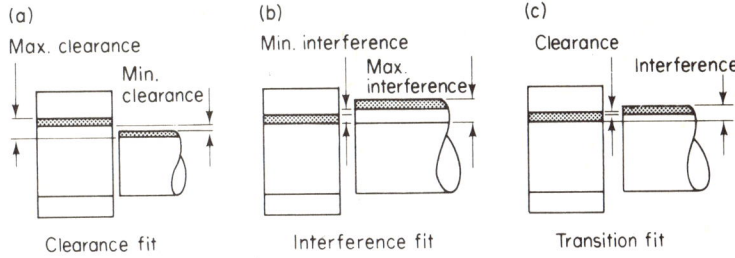

Figure 3.43 Types of fit

An *interference fit* will always be produced if the low limit on one part is greater than the high limit on the mating part. The allowance on this type of fit will always be negative, which means that force will have to be applied to assemble the two parts together.

A *transition fit* is one whereby, depending on the size of the two parts, a clearance or interference fit may be produced. The allowance could therefore be positive or negative. A transition fit is specified where a light push fit is required.

Example: A hole and shaft are specified as follows:

$$\text{Hole } 46.50 \begin{array}{l} +0.05 \\ +0.02 \end{array} \text{mm} \quad \text{Shaft } 46.50 \begin{array}{l} +0.01 \\ +0.00 \end{array} \text{mm}$$

Determine: (a) The limits of size for hole and shaft
(b) The tolerances on the hole and shaft
(c) The maximum and minimum allowances.

(a) Limits of hole size = $\dfrac{46.55}{46.52}$ mm

Limits of shaft size = $\dfrac{46.51}{46.50}$ mm

(b) Hole tolerance = $46.55 - 46.52 = 0.03$ mm
Shaft tolerance = $46.51 - 46.50 = 0.01$ mm

(c) Maximum metal condition (hole)
$= 46.52$ mm
Maximum metal condition (shaft)
$= 46.51$ mm
Minimum allowance
$= 46.52 - 46.51 = 0.01$ mm
Minimum metal condition (hole)
$= 46.55$ mm
Minimum metal condition (shaft)
$= 46.50$ mm
Maximum allowance
$= 46.55 - 46.50 = 0.05$ mm

LIMITS-AND-FITS SYSTEMS

The purpose of a *limits-and-fits system* is to provide a range of tolerances for different classes of fit. Such systems may be hole- or shaft-based. A *hole-based system* is one where the hole limits of size are kept constant and the shaft varied in size to provide the required class of fit. A *shaft-based system* is one where the shaft limits of size are kept constant and the size of the hole varied to provide the required class of fit. Users of limits-and-fits systems generally employ the hole-based system since it is relatively easy to standardise the size of the hole by using standard tooling such as drills and reamers.

Fundamental deviation

The term *fundamental deviation* is used to denote the position of the tolerance band on both shaft and hole relative to the basic size. The fundamental deviation therefore determines the class of fit, i.e. clearance, interference, or transition.

The fundamental deviation for a clearance fit is illustrated in Figure 3.44.

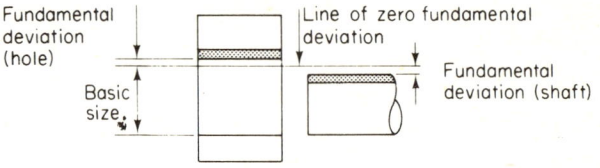

Figure 3.44 Fundamental deviation

BS 4500: 1969 (ISO limits and fits)

In the past, many systems for limits and fits have been devised. The one now generally used and accepted is BS 4500 which is based on the system adopted by the International Organisation for Standards (ISO). The system covers a wide range of basic sizes up to 3150 mm. BS 4500 provides for 28 fundamental deviations for both holes and shafts, the holes being designated by a capital letter and the shafts by a small letter, as shown:

Holes: A, B, C, CD, D, E, EF, F, FG, G, H, JS, J, K, M, N, P, R, S, T, U, V, X, Y, Z, ZA, ZB, ZC
Shafts: a, b, c, cd, d, e, ef, f, fg, g, h, js, j, k, m, n, p, r, s, t, u, v, x, y, z, za, zb, zc

Each of these fundamental deviations is further provided with 18 grades of tolerance, each grade being identified by the notation ITO1, ITO, IT1, IT2, IT3 to IT16. The tolerance grade is related to the basic size, i.e. the larger the basic size the

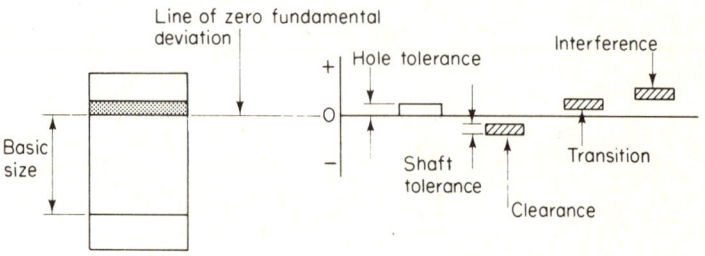

Figure 3.45 BS 4500: 1969 hole-based system (H series) (BSI)

larger the tolerance that must be applied. It will no doubt be apparent to the reader that BS 4500 provides an extremely large number of tolerances to suit different classes of fit. However, for the majority of engineering applications the H hole is used which has a fundamental deviation equal to zero. This means that the basic size equals the low limit of the hole. This is illustrated in Figure 3.45 together with a tolerance diagram showing the different classes of fit.

Specifying types of fit When specifying a type of fit the fundamental deviations of the hole and shaft must be stated together with their respective tolerance grades. For example, a clearance fit is specified as being 14 mm H7g6. The basic size on both parts will be 14 mm diameter. The designation H7 means that the hole has an H fundamental deviation with a tolerance grade of 7, while g6 means the shaft has a g fundamental deviation with a tolerance grade of 6. To obtain the actual tolerance for specified fits, reference must be made to tolerance tables as laid down by BS 4500:1969, of which a small part has been extracted (Figure 3.46) as an example of each class of fit.

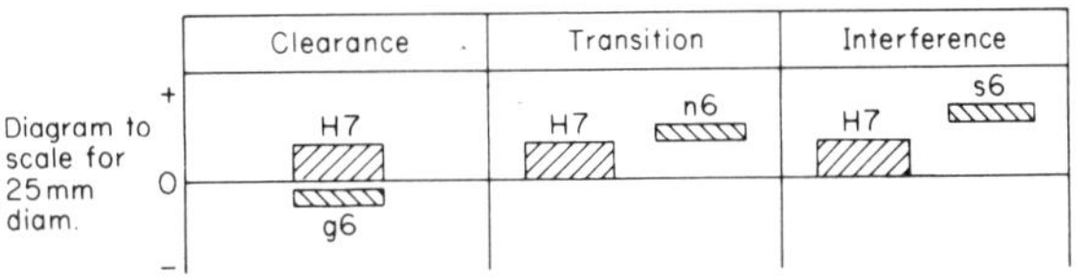

Nominal sizes		Tolerance		Tolerance		Tolerance	
Over	To	H7	g6	H7	n6	H7	s6
mm	mm	0.001 mm	0.001 mm	0.001 mm	0.001 mm	0.001 mm	0.001 mm
–	3	+10 / 0	-2 / -8	+10 / 0	+10 / +4	+10 / 0	+20 / +14
3	6	+12 / 0	-4 / -12	+12 / 0	+16 / +8	+12 / 0	+27 / +19
6	10	+15 / 0	-5 / -14	+15 / 0	+19 / +10	+15 / 0	+32 / +23
10	18	+18 / 0	-6 / -17	+18 / 0	+23 / +12	+18 / 0	+39 / +28
18	30	+21 / 0	-7 / -20	+21 / 0	+28 / +15	+21 / 0	+48 / +35

Figure 3.46 BS 4500: 1969 limits and fits (hole-based) (BSI)

Referring to our previous example of 14 mm H7g6 the tolerance from the table will therefore be:

Hole: $14.000 \, {}^{+0.018}_{+0.000}$ mm   Shaft: $14.000 \, {}^{-0.006}_{-0.017}$ mm

The limits of size for the hole will be $\frac{14.018}{14.000}$ and the shaft will be $\frac{13.994}{13.983}$ mm.

For this fit, the maximum and minimum clearance will therefore be 0.035 mm and 0.006 mm respectively.

LIMIT GAUGING

When components are manufactured in large quantities it is much quicker and hence more economic to inspect these by *limit gauging* rather than by direct measurement. Limit gauging is a technique designed to indicate to the operator or inspector as to whether the component is inside or outside the specified limits of size. The use of limit gauging also lends itself to the use of semi-skilled operators.

A limit gauge generally consists of two gauging elements known as the GO and NOT GO. One of the most common limit gauges in general use is the *plug gauge* (Figure 3.47) used for gauging the diameter of a hole. The function of the GO end on this gauge is to ensure that the hole is not smaller than its low limit, while the NOT GO is used to ensure that the hole is not larger than its high limit. Therefore, provided that the GO gauge enters the hole and the NOT GO does not then the hole must be within the specified limits of size.

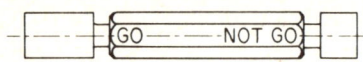

Figure 3.47 Plug gauge

Taylor's principle of gauging

Although there are many different types of limit gauge, they should all be designed in accordance with Taylor's principle, which states:

(a) The GO gauge checks the maximum metal condition (low limit for a hole, high limit for a shaft) and should also check as many dimensions as possible within the feature to be gauged.

(b) The NOT GO gauge checks the minimum metal condition (high limit for a hole, low limit for a shaft) and should check only one dimension within the feature to be gauged.

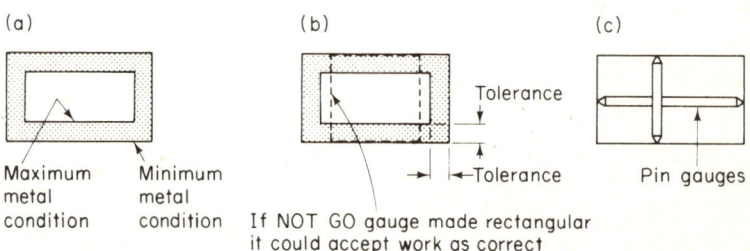

Figure 3.48 Taylor's principle of gauging

To understand this principle more fully consider the rectangular hole illustrated in Figure 3.48(a). To comply with Taylor's principle the GO gauge must be made rectangular and such that its dimensions conform to the low limit on both length and width. If, however, the NOT GO is also made rectangular and to the high limit then a condition shown in Figure 3.48(b) could exist. Here the gauge will not enter the hole and will therefore accept the work as being correct even though the length is outside specified limits. This means then, that to gauge the length and width two separate gauges must be used, so that each will gauge only one dimension. These gauges usually consist of pin gauges as shown in Figure 3.48(c).

Gauge tolerances and wear allowance

When specifying sizes for limit gauges consideration must be given to *gauge tolerances*, since we cannot expect the gaugemaker to make the gauge exactly to size. As a general rule the gaugemaker's tolerance is quoted as 10% of the work tolerance and is disposed relative to the basic size opposite to the direction that the gauge wears.

To overcome the initial wear on limit gauges a 'wear allowance' is sometimes applied to the GO gauge which is generally specified as 20% of the gaugemaker's tolerance. If the work tolerance is less than 0.10 mm, the wear allowance is usually ignored since it becomes too small to be of any practical significance. The wear allowance is never applied to the NOT GO gauge since this should never wear, i.e. a NOT GO plug gauge should not enter a hole.

The disposition of gauge tolerances and the wear allowance for plug gauges and ring or gap gauges is shown in Figure 3.49.

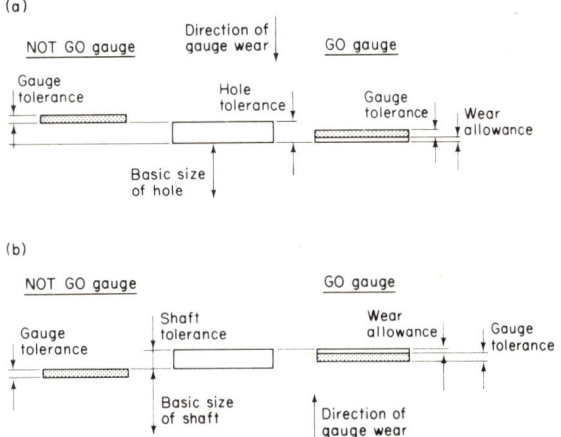

Figure 3.49 Gauge tolerances, (a) disposition of tolerances for GO, NOT GO plug gauges, (b) disposition of tolerances for GO, NOT GO ring and gap gauges

Example: A hole is specified as $15.00 ^{+0.06}_{+0.02}$ mm diameter. Determine the gauge limits for a GO, NOT GO plug gauge to check this dimension.

Maximum metal condition = 15.02 mm
Minimum metal condition = 15.06 mm

Therefore work tolerance = 15.06 − 15.02 = 0.04 mm

Gaugemaker's tolerance = 10% of 0.04 mm
     = **0.004 mm**

The limits on the GO and NOT GO gauge will therefore be

GO gauge
High limit = 15.024 mm
Low limit = 15.020 mm

NOT GO gauge
High limit = 15.064 mm
Low limit = 15.060 mm

Types of limit gauge

In addition to the plug gauge already described there are many types of limit gauge in general use designed to check different workpiece features.

The diameter of external and internal tapers may be checked using *taper plug gauges* and *taper ring gauges*, as shown in Figure 3.50. The high and low limits on these gauges are represented by a ground step, so that provided the work surface lies within this step the work will be within the specified tolerance.

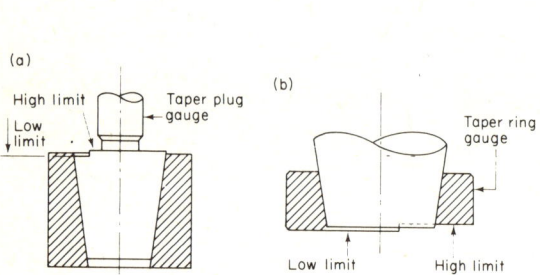

Figure 3.50 Gauging external and internal tapers

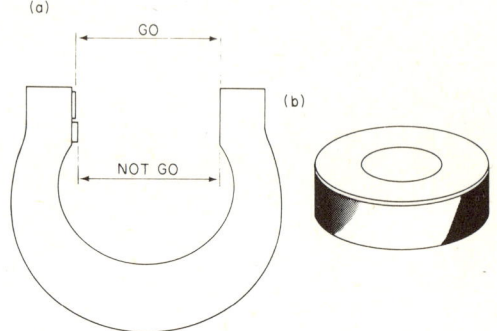

Figure 3.51 Gauges for external diameters, (a) caliper gauge, (b) ring gauge

Gauges for checking diameters usually consist of *caliper gauges* and *ring gauges*, as shown in Figure 3.51. The caliper gauge may be of the fixed or adjustable type. The adjustable type has the advantage that the anvils may be adjusted to suit a range of sizes, the gauging dimension being set using slip gauges.

SUMMARY

Comparators are used in precision measurement for comparing the size of the workpiece with that of a known standard, e.g. slip gauge. The magnification systems used in comparators are based on mechanical, electrical, optical or pneumatic principles.

Optical projectors are used for measuring an enlarged projected image of the workpiece. Linear and angular dimensions may be checked by this method together with geometric form.

The autocollimator, angle dekkor and spirit level are instruments for measuring small angular displacements.

Alignment tests are carried out on machine tools to determine the accuracy of the geometrical relationships between the various moving elements. The equipment used consists mainly of test mandrels and dial indicators.

Measurement of surface finish is specified by the arithmetical mean deviation (R_a). Assessment of surface finish may be carried out by comparing the workpiece finish with a standard finish or using an instrument employing a stylus.

A system of limits and fits is used in engineering manufacture to promote interchangeability, uniform quality, and economic manufacture.

Components produced in large quantities are usually inspected by limit gauging. The design of limit gauges should conform to Taylor's principle of gauging.

QUESTIONS

(1) Using a suitable sketch, explain the principle of comparative measurement.

(2) With the aid of a clear diagram explain the operating principle of either

      (i) Mechanical comparator
  OR (ii) Pneumatic comparator

(3) List FOUR desirable features that a comparator should possess.

(4) By means of a clearly-labelled diagram show the optical system of a projector suitable for engineering inspection. Explain the importance of using collimated light for optical projection.

(5) Using a sketch, explain what adjustment is required to a projector when projecting helical features such as screw threads.

(6) State THREE common magnifications used in optical projection.

(7) Explain with the aid of suitable diagrams the operating principle of the autocollimator. State two applications of this instrument.

(8) Describe, using a sketch, how the component shown in Figure 3.52 may be measured using an angle dekkor.

(9) Describe, using a sketch, the 'field of view' as seen in an angle dekkor.

(10) Explain, using suitable diagrams, how the following machine tool alignments may be carried out:

  (a) Spindle axis square with table on a drilling machine.
  (b) Tailstock quill movement parallel with bed on a centre lathe.

(11) Using clear diagrams, show what is meant by the following types of fit:

    (i) Clearance (ii) Interference (iii) Transition.

(12) An interference fit for a hole and shaft is specified as 8 mm H7s6. Using the table in Figure 3.46 determine:

  (a) The maximum and minimum metal condition for both hole and shaft.
  (b) The maximum and minimum interference between the hole and shaft.

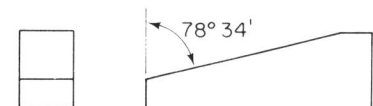

Figure 3.52

(13) State Taylor's principle of gauging.

(14) By means of a clear diagram show the disposition of tolerances and the wear allowance for a general purpose GO, NOT GO, plug gauge.

(15) A shaft is specified as $24.00 \begin{smallmatrix} +0.08 \\ +0.02 \end{smallmatrix}$ mm diameter. Determine the gauge limits for a GO, NOT GO caliper gauge to check this dimension.

4 Machining

INTRODUCTION Many types of machining processes exist, but most may be regarded as those employing either *single-* or *multi-point cutting tools*. Single-point tools are those found in turning, shaping and planing operations, while multi-point cutting tools include those found in milling and grinding processes. Since engineering manufacture must be regarded as an economic activity a great deal of research has been undertaken to develop tools that will have a long tool life and yet will perform efficiently when cutting. The purpose of this chapter, therefore, is to outline the use of single- and multi-point tools and also to see how they may be used for specific machining operations.

SINGLE-POINT TOOLS Readers will already be familiar with the basic theory of metal cutting from their studies at level 2. This section is intended to develop this previous work into some of the more important aspects of single point tooling.

Forces at the tool point During metal cutting with single-point tools there are three forces which act at the tool point, each force acting mutually perpendicular to each other. Figure 4.1(a) shows these forces

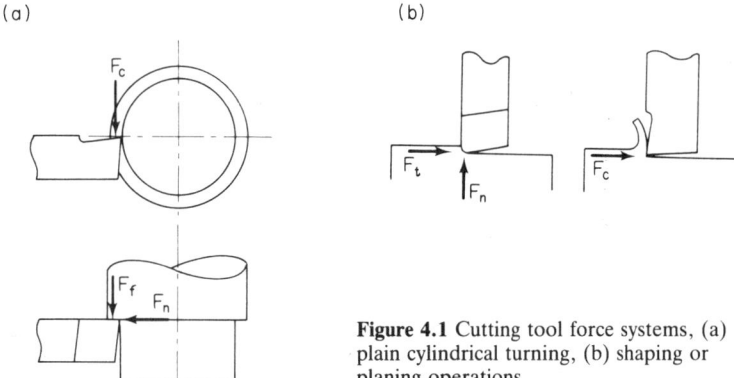

Figure 4.1 Cutting tool force systems, (a) plain cylindrical turning, (b) shaping or planing operations

acting at the tip of a single-point tool during plain cylindrical turning.

F_c is known as the *cutting force* and acts vertically to the tool face.
F_f is known as the *feed force* and acts parallel and horizontally to the work axis.
F_n is known as the *reaction between tool and work* and is horizontal and at 90° to the work axis.

Of the three forces F_c and F_f are the largest, F_c being the most important force regarding power consumption.

Although Figure 4.1(a) depicts the tool force system during a turning operation, the same principle applies to the shaping of a plane surface. The only difference being that the feed force (F_f) is replaced by a side thrust (F_t), as illustrated in Figure 4.1(b).

TOOL-FORCE MEASUREMENT

To measure cutting tool forces a *dynamometer* is used, which is an instrument designed to detect small deflections of the cutting tool when it is subjected to the forces already described. There are many different types of dynamometer embodying different principles. The main design features are that they should be as rigid as possible so as not to be influenced by machine tool vibrations, and also have sufficient sensitivity to detect small displacements of the cutting tool. Figure 4.2. shows a simple dynamometer for use on a centre lathe, suitable for measuring the three forces F_c, F_f and F_n. When measuring cutting tool forces in practice the force F_n is often ignored, since it is too small to be of any significance.

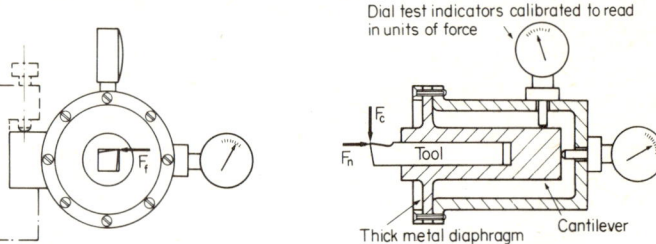

Figure 4.2 Lathe tool dynamometer

During cutting, the cutting and feed forces cause the thick metal diaphragm to deflect. These deflections are magnified by the cantilever whose displacements are recorded by the dial indicators which are calibrated to read in units of force.

The results obtained from a tool dynamometer may be used to calculate the power consumed during cutting.

POWER CONSUMPTION

During any metal cutting operation the power supplied to the machine tool is not used solely for cutting, in fact most of the power supplied is required to drive the machine itself. The power required for cutting is related to the magnitude of the forces at the tool point, of which F_c is the most significant. The forces F_c and F_n are generally regarded as being too small to be of any practical value in power calculations.

The conditions existing during a machining operation on a centre lathe are illustrated in Figure 4.3.

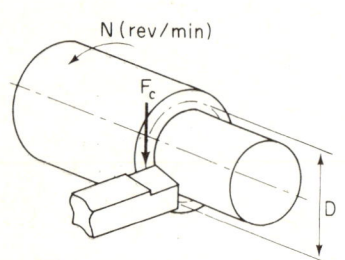

Figure 4.3 Requirements for power calculations

$$\text{Power} = \frac{\text{Work done (J)}}{\text{Time taken (s)}} = \text{J/s} = \text{watts}$$

Note that work done = Force × Distance = newton metres = joules (J)

Work done due to cutting force for one revolution

$$= F_c \times \frac{\pi D}{1000}$$

Therefore,

$$\text{Power due to cutting force} = F_c \times \frac{\pi DN}{1000 \times 60} \text{ watts}$$

where D = mean diameter in mm
N = spindle speed in rev/min.

Example: During a cutting test using a tool dynamometer on a centre lathe, the following results were recorded:

Spindle speed 450 rev/min   Vertical cutting force 1500 N
Blank diameter of work 50 mm   Depth of cut 4 mm

Determine the power consumed in cutting during this test.

$$\text{Power due to cutting} = F_c \times \frac{\pi DN}{1000 \times 60}$$
$$= 1500 \times \frac{\pi \times 46 \times 450}{1000 \times 60} = \frac{97546}{60}$$
$$= 1625.7 \text{ W}$$
Say,   **1.63 Kw**

METAL REMOVAL RATES

During a machining process manufacturing engineers are interested, from the point of view of economy, in the rate at which metal can be removed from the workpiece. This does not necessarily mean that it is more economic to machine using fast speeds and feeds, since this condition would produce high power consumption and promote rapid tool wear, leading to expensive re-grinding time and tool replacement. Therefore, engineers must optimise and consider a metal removal rate which will be both efficient and economic, taking into account relevant factors such as speed, feed, depth of cut and surface finish of the workpiece.

In general the metal removal rate is the amount of metal removed in one minute and is often expressed in mm^3/min. Therefore,

$$\text{Metal removal rate} = \frac{\text{Volume (mm}^3)}{\text{Time (min)}}$$

For a plain cylindrical turning operation the metal removal rate is given by:

Cutting speed (mm/min) × Feed (mm/rev) × Depth of cut(mm)

TOOL LIFE

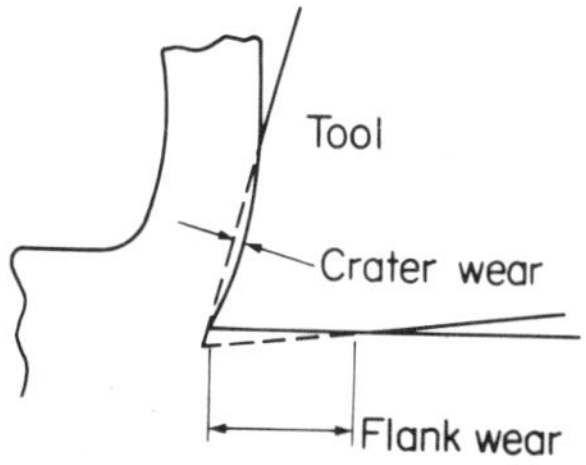

Since the life of a cutting tool is directly related to the amount of wear that takes place at the tool point, some consideration must be given as to how a tool wears. During cutting two forms of wear are produced, *flank wear* at the tip of the tool and *crater wear* behind the tip caused by the underside of the chip rubbing against the top of the tool. Both forms of wear are illustrated in Figure 4.4. Flank wear is the most significant since this will cause the work to be machined oversize when pre-set tools are used. Furthermore, flank wear will also bring about a deterioration in surface finish.

Figure 4.4 Forms of tool wear

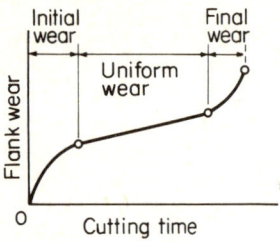

Figure 4.5 Relationship between cutting time and flank wear

If a tool is allowed to cut continuously until it fails completely, then it can be shown that the life of the tool can be divided into three distinct stages, as illustrated in Figure 4.5.

1st stage: The initial breakdown period when the tool wears rapidly and a small wear land is established.
2nd stage: The region of uniform wear.
3rd stage: The final and rapid breakdown period, when the tool wear rate increases rapidly until it finally fails.

In practice the tool would be re-ground before the third stage is reached.

A great deal of investigation has been carried out in an attempt to predict the expected life of a cutting tool between re-grinds. A pioneer in this field was F. W. Taylor, who established the well-known expression

$$VT^n = C$$

where V = cutting speed in m/min
T = expected tool life between re-grinds in minutes
n = a constant which is related to the cutting tool material used, of which the following are examples:

High speed steel	0.1 to 0.15
Tungsten carbide	0.2 to 0.4
Ceramics	0.4 to 0.6

C = a constant, dependent upon cutting conditions, e.g. depth of cut, feed and tool geometry.

The application of this formula should be exercised with care, since it is based on experimental results, and will hold true only for a given set of specified cutting conditions. Under these conditions, the only values permitted to change are V and T.

The following example will demonstrate how this formula may be applied.

Example: When machining mild steel at 30 m/min using a high speed steel tool it was found that the tool had to be re-ground every three hours. Calculate the expected life of this tool when used under the same conditions, but using a cutting speed of 40 m/min. Assume the constant n to have a value of 0.12

$$VT^n = C$$

Since T is raised to the power n, this expression may be rewritten as

$\log C = \log V + n \log T$
$ = \log 30 + 0.12 \log 180$
$ = 1.4771 + (0.12 \times 2.2553)$
$ = 1.7477$
Antilog $1.7477 = 55.94$

Therefore, $C = 55.94$
$ \log 55.94 = \log 40 + 0.12 \log T$
$ 0.12 \log T = \log 55.94 - \log 40$
$ 0.12 \log T = 1.7477 - 1.6021$

Therefore, $\log T = \dfrac{0.1456}{0.12} = 1.2133$

Antilog $1.2133 = 16.34$

Expected tool life $T = 16.34$ min   Say, **16 min.**

It should be understood that the time T is the actual time while cutting and not the time that the tool is in the machine.

SINGLE-POINT TOOL CONSTRUCTION

Although high speed steel is a popular cutting tool material, modern machining methods employ carbide and ceramic tooling. Since these tool materials are extremely hard they are also brittle and that means they require a special type of toolholder.

Cemented carbide tipped tools

Nowadays these tools fall into two main groups:

(i) Brazed tipped tools,
(ii) Tools using indexable or throw-away tips.

Brazed tipped tools

These tools are used extensively in production and are commonly used in shaping, planing and turning operations. Many shapes and sizes are available with different tool angles to suit different cutting conditions. They are produced by machining a recess into the tip of a medium carbon steel shank and then brazing in the carbide insert. The brazing material is in the form of a shim, which, when it melts, flows by capillary action between the shank and insert. So that heating is uniform and the risk of scaling eliminated, a high-frequency induction heating coil is often used, as shown in Figure 4.6.

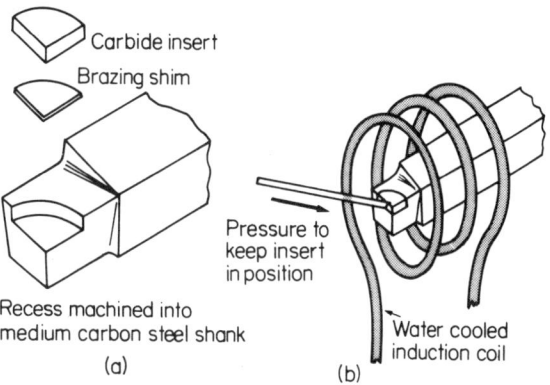

Figure 4.6 Preparation of a carbide-tipped tool, (a) components, (b) tool is heated by eddy currents induced as a result of magnetic field produced by high-frequency current in coil

Indexable or throw-away tips

These tools are rapidly becoming popular and offer the following advantages compared with brazed tipped tools.

(a) The design of the tool holder allows a number of cutting edges to be used on the same tool, simply indexing the tip to the next position when one edge has become dulled.

When using negative rake a three- or four-sided tip may be turned over thereby providing six or eight cutting edges respectively.

(b) After all edges have been used, the tip is thrown away and replaced by another. Taking into account the initial cost of the tip, this is more economical since the non-productive time/cost of re-grinding a brazed tipped tool would be greater, an important consideration in batch and volume production.

(c) Chip breakers may be built into the toolholder, which may be adjusted to suit metal-cutting conditions.

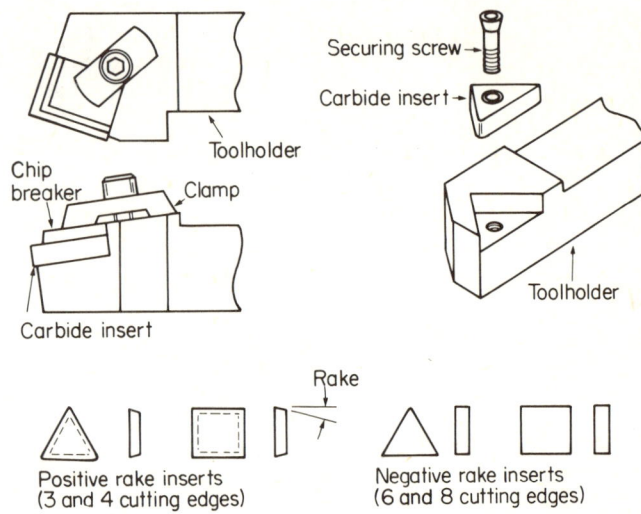

Figure 4.7 Throw-away tip toolholders and inserts

Due to the difficulty of brazing ceramics into steel shanks, they are always used in the form of throw-away tips. Figure 4.7 shows the construction of typical throw-away tip toolholders and the types of insert available.

Quick-change and pre-set tooling

When tools are used to manufacture large quantities it is important to keep re-grinding and tool setting time to a minimum, so that the machining process remains economic. These times may be minimised by using pre-set tools in conjunction with quick-change toolholders, so that a duplicate tool can be substituted when it is necessary to re-grind an existing tool.

The design of *quick-change toolholders* varies, but should be such that they can be fitted easily, and also give precise repeatability of position. This is important if the dimensions of the machined work are governed by stops to control the movement of the machine slides. Figure 4.8 shows a typical arrangement for a quick-change toolholder.

When duplicate tools are used, then these must be set using a *pre-set tooling fixture* so as to ensure repeatability when the tool is fitted to the machine. Figure 4.9 shows a typical fixture used in conjunction with dial test indicators.

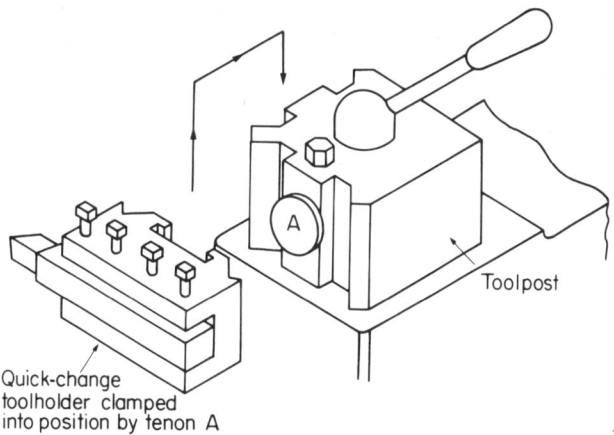

Figure 4.8 Quick-change tooling

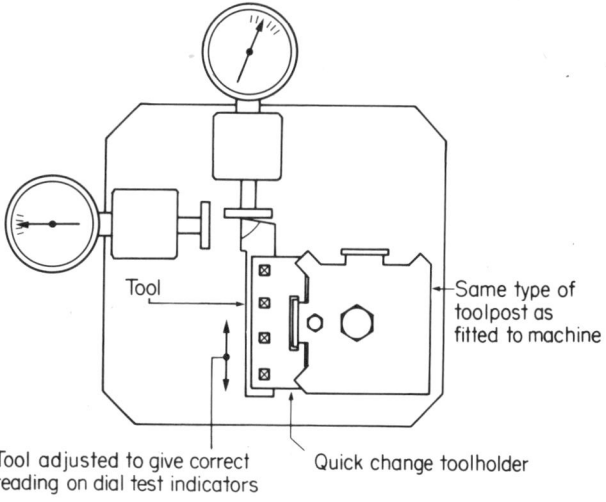

Figure 4.9 Pre-set tooling fixture

MILLING Milling is generally recognised as being one of the most important machining processes, on account of its versatility and high metal-removal rate. The process consists basically of producing flat or profiled surfaces by means of a rotating multi-tooth cutter.

TYPES OF CUTTER There are many cutters of different shapes and sizes designed for use on either horizontal or vertical spindle machines. However, these cutters may be conveniently classified into three groups, these being

(i) Fluted cutters
(ii) Form-relieved cutters
(iii) Inserted-tooth cutters.

Fluted cutters

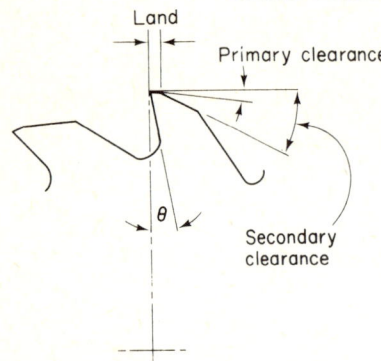

Figure 4.10 Fluted cutter geometry

The teeth of these cutters are designed to cut on the periphery and in many cases on the side as well. The essential geometry of cutters with fluted teeth is illustrated in Figure 4.10.

A rake angle θ is provided on the front of the tooth and usually has a value of about 12°, being satisfactory for a wide range of ferrous materials. For non-ferrous materials a rake angle of between 20–25° is generally required. A primary clearance must be provided on the top of the tooth extending over a small width known as the 'land'. The width of this land will depend on the size of the cutter, but generally will be in the order of about 1–2 mm, the primary clearance having a value of about 4°. The secondary clearance should be kept as small as possible so as to provide adequate strength to the tooth, while at the same time providing sufficient space for the chips.

Fluted cutters are always sharpened by grinding the land.

SLAB-MILLING CUTTER

This cutter (Figure 4.11) is designed for producing flat surfaces or where large stock removal is required. The teeth may be parallel to the cutter axis or cut to form a helix. Cutters with straight teeth have a tendency to cause chatter since the teeth make contact across the whole width of the work. For this reason most slab cutters have helical teeth which means that after a tooth has made contact with the work the chip will gradually increase in width. Furthermore, the helix angle permits several teeth to cut simultaneously which results in a much smoother cutter action.

Figure 4.11 Slab-milling cutter

SIDE-AND-FACE CUTTER

This cutter (Figure 4.12(a)) is considerably narrower than the slab cutter and is generally of a larger diameter. As well as cutting on the periphery, the side-and-face cutter is also designed to cut on the sides, thus making it suitable for machining slots and vertical faces, as shown in Figure 4.12(b).

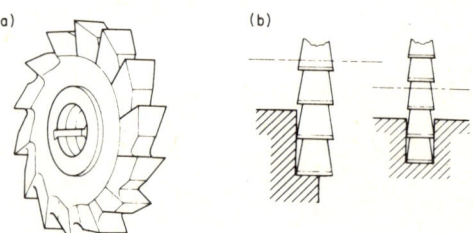

Figure 4.12 Side-and-face cutter

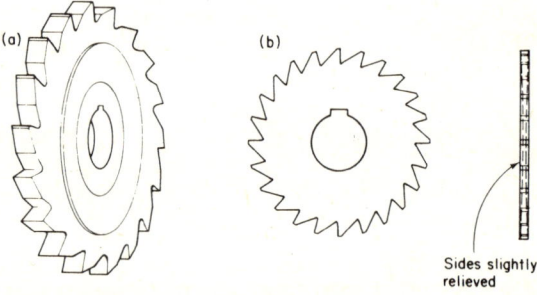

Figure 4.13 (a) Slotting cutter, (b) slitting saw

Machining 71

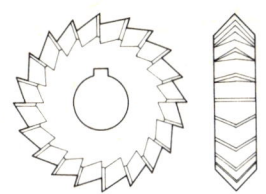

Figure 4.14 Angle cutter

Figure 4.15 End mill

SLOTTING CUTTER
This cutter has teeth only on the periphery and is used solely for cutting slots. Where narrow slots are required slitting saws are used in addition to normal sawing operations. Examples of a slotting cutter and slitting saw are illustrated in Figure 4.13.

ANGLE CUTTER
This cutter (Figure 4.14) is used for machining angles and Vs and may be either single- or double-angle. Typical angles are 30°, 45°, 60° and 90°.

END MILL
This type of cutter (Figure 4.15) is used mainly on vertical milling machines and is designed to cut on the front and side, thus making it suitable for machining steps and slots. The diameters of end mills vary from about 2 mm up to about 35 mm. The majority of end mills have straight shanks which have to be held in a collet-type chuck, while some have a morse taper shank which fits directly into the machine spindle.

SHELL MILL
This cutter is larger in diameter compared with the end mill and is generally used for machining large flat surfaces. Unlike end mills, shell mills require a special arbor. This is provided with a parallel diameter to locate the cutter and tenons to provide a positive drive, the cutter being clamped by means of a screw. The shell mill and its arbor are illustrated in Figure 4.16.

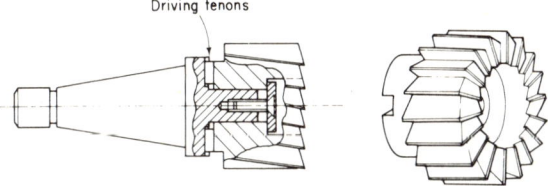

Figure 4.16 Shell mill and arbor

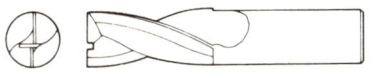

Figure 4.17 Slot drill

SLOT DRILL
This cutter is used to machine slots to a specified depth, e.g. a keyway in a shaft. The slot drill has only two flutes and it will be found that the front cutting edge of one flute is slightly longer than the other. This permits the cutter to be fed axially into the work before the longitudinal feed is applied, an operation that cannot be done with an end mill due to the recess found on the front of this cutter. A typical slot drill is illustrated in Figure 4.17.

T-SLOT AND DOVETAIL CUTTER
The T-slot cutter (Figure 4.18(a)) is used for producing slots such as those found in a milling machine table. Before this cutter can be used, a slot must first be machined with an end mill to provide clearance for the parallel shank of the cutter.

The dovetail cutter (Figure 4.18(b)) is used for machining internal angular surfaces such as those found on machine tool slides. Dovetail cutters usually have angles of 30° or 60°.

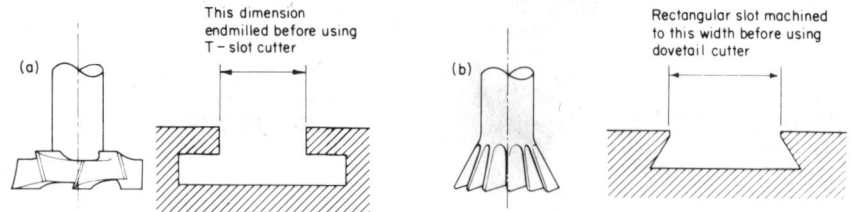

Figure 4.18 (a) T-slot cutter, (b) dovetail cutter

Form-relieved cutters   These cutters are used to impart a specified profile onto the workpiece. The geometry of form-relieved cutters is such that their form is always maintained irrespective of the number of times that they are sharpened. This is achieved by grinding the front face of each tooth radial to the centre of the cutter, as shown in Figure 4.19(a).

There are many different types of form-relieved cutter, some of the more common being shown in Figure 4.19(b).

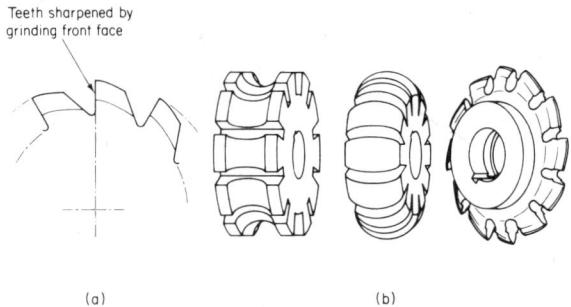

Figure 4.19 Form-relieved cutters, (a) sharpening a cutter, (b) cutters, left to right, convex, concave and involute gear

Inserted-tooth cutters   Cutters in this group are confined mainly to large-diameter cutters known as *face mills*. These are used for machining large, flat surfaces on either horizontal or vertical spindle machines. The body of an inserted-tooth cutter is made from an alloy steel and has the facility of locating and clamping teeth in the form of inserts. These inserts may be made from high-speed steel or tungsten carbide. Two methods of securing the inserts are illustrated in Figure 4.20. In Figure 4.20(a) the insert is held in position by means of a wedged clamp. After every tooth has been clamped the cutter must then be ground to ensure that

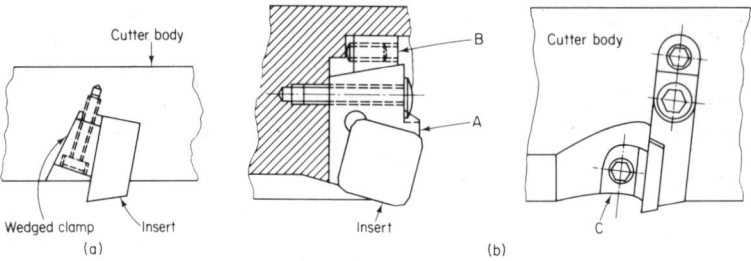

Figure 4.20 Inserted-tooth cutters (Kennametal Ltd)

Machining

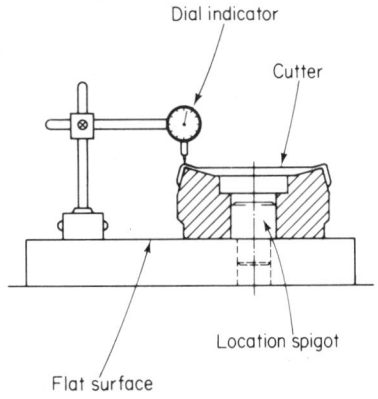

Figure 4.21 Inserted-tooth cutter setting fixture (Kennametal Ltd)

all the inserts are at the same weight. The insert arrangement shown in Figure 4.20(b) is of the throw-away type. Here the insert is located in a detachable shoe *A* which can be adjusted axially by means of the wedge *B*, the insert assembly being secured by means of the clamp *C*.

To ensure that the inserts shown in Figure 4.20 are set to the correct height a special fixture (Figure 4.21) is used whereby each insert is adjusted until the same dial indicator reading is obtained over the tip of each insert. It is common practice with this type of cutter to have one insert slightly higher than the others. This is often known as a 'scraper' and it assists in improving the finish of the workpiece.

CUTTING TECHNIQUES

There are two ways in which the work may be presented to the cutter, these being known as

(i) Up-cut milling (ii) Down-cut milling.

Both methods are illustrated in Figure 4.22.

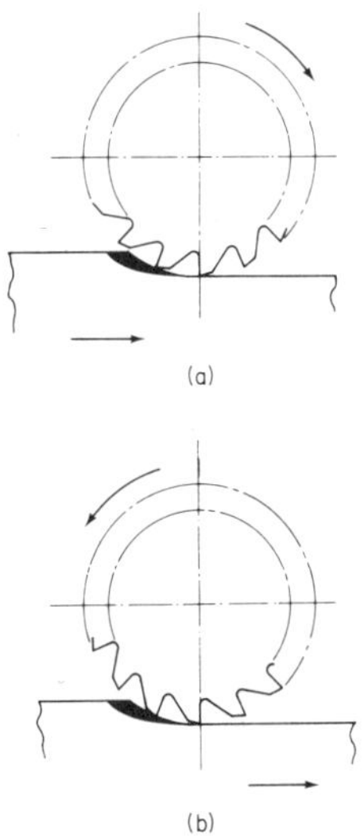

Conventional practice utilises the *up-cut milling* method whereby the work is presented to the cutter in such a way that the chip produced increases to a maximum thickness after starting at the full depth of cut. Although this method represents the most common practice it does have the disadvantage that the cutter has a tendency to lift the work. Furthermore, since the work feed opposes the cutter rotation, this means that in addition to the power required for cutting, extra power must be supplied to the machine spindle to overcome the feed force.

In *down-cut milling*, sometimes known as *climb milling*, the work is fed into the cutter so that the chip starts at a maximum thickness at the surface of the work. In this way the chip thickness will reduce to zero at the full depth of cut. Due to the table feed being in the same direction as the cutter the power consumption will be less compared with up-cut milling since no additional power is required to overcome any feed force. Furthermore, the cutter rotation also assists in keeping the work against the machine table.

Owing to the tendency for the work to be dragged beneath the cutter, it is essential that machines used for down-cut milling are equipped with a *backlash eliminator*, a device designed to ensure that there is no free movement between the machine leadscrew and nut. It is also important when using down-cut milling that the machine table slideway is in a proper state of adjustment.

Figure 4.22 (a) Up-cut milling, (b) down-cut (climb) milling

STRADDLE AND GANG MILLING

Some components machined by milling require two faces to be parallel to each other. On machines used for production milling this operation is often performed by *straddle milling*. This consists of using two side-and-face cutters set to the required distance on the arbor by means of spacing collars, as shown in Figure 4.23(a).

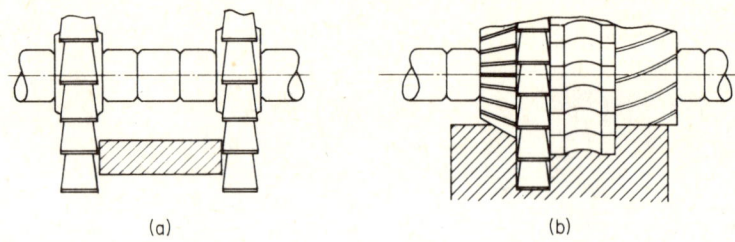

Figure 4.23 (a) Straddle milling, (b) gang milling

When a surface needs to be cut with a series of shapes, e.g. steps, angles, curves, etc., these may be machined simultaneously by *gang milling*. This involves feeding the work beneath several cutters whose diameters, lengths and shapes are controlled in such a way that they will form the workpiece into the required shape. Since the length of cut is considerably increased due to the use of several cutters, the cutting force and hence power will also increase. This means that care should be exercised as to the power availability of the machine, especially if a heavy cut is to be taken. Another important consideration when gang milling is that additional arbor support brackets may have to be used to ensure that arbor deflection is kept to a minimum. A typical gang milling operation is shown in Figure 4.23(b).

CUTTING SPEEDS AND FEEDS

Cutting speed for milling corresponds to the peripheral speed of the cutter measured in metres/minute and is chosen according to the material to be machined. Some typical values are shown below.

Material	Cutting speed m/min
Cast iron and mild steel	15–35
High carbon steel	10–20
Brass	50–60
Aluminium	25–200

It should be appreciated that the values quoted are meant to serve only as a guide, since cutting speeds are influenced by the cutter material and, to a large extent, by the condition of the machine. In order to obtain the required cutting speed, the correct spindle speed in revolutions per minute must be determined, and this may be found from the following expression:

$$N = \frac{1000S}{\pi D}$$ where N = spindle speed (rev/min)
S = cutting speed (m/min)
D = cutter diameter (mm)

Feeds for milling are usually quoted as the speed of the machine table in mm/minute and are chosen according to the metal-removal rate required or surface finish requirements. Since milling cutters vary in terms of their number of teeth, diameter and type it is necessary to specify the feed per tooth before the table feed can be determined. The table feed (F)

may be found from the following expression:

$F = N \times n \times f$ where N = spindle speed (rev/min)
n = number of teeth in cutter
f = feed per tooth (mm)

Example: A 65 mm-diameter slab mill having eight teeth is used to take a finishing cut on a mild steel workpiece. If the cutting speed used is 28 m/min and the feed per tooth is 0.05 mm, determine the machine table feed in mm/min.

Spindle speed $N = \dfrac{1000S}{\pi D} = \dfrac{1000 \times 28}{\pi \times 65}$

$= \dfrac{28000}{204.2} = 137$ rev/min

Table feed $F = N \times n \times f$
$= 137 \times 8 \times 0.05$
$= 54.8$ Say, **55 mm/min**

It should be appreciated that milling machines have only a limited range of feed rates, which means that the nearest available feed will have to be chosen.

MILLING MACHINE ACCESSORIES

The versatility of milling machines is greatly extended by the use of equipment such as the *dividing head* and the *rotary table*.

The dividing head

When features such as splines, gear teeth, etc. have to be cut around the circumference of a disc a dividing head is used. This is a device designed to rotate (or index) the work by a specified amount so that the work is positioned correctly relative to the cutter. Also, when used in conjunction with suitable gearing, as will be shown later, the dividing head may be used for machining helical forms such as the fluting found on twist drills, reamers and helical milling cutters.

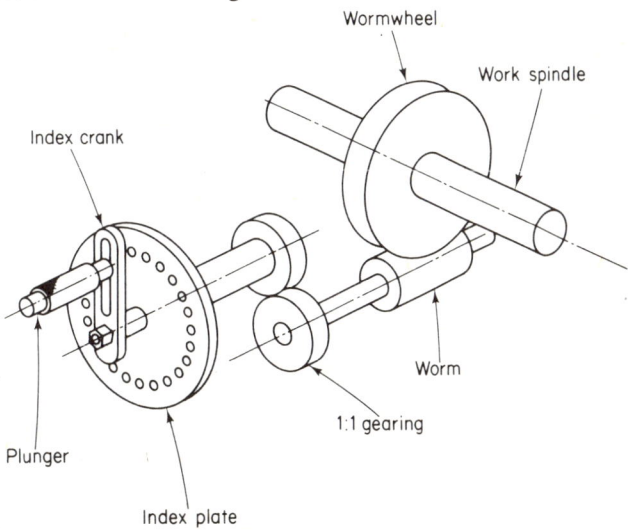

Figure 4.24 Dividing head

Basically, the dividing head consists of a 40-tooth worm wheel fitted to the work spindle which meshes with a single-start worm, as shown in Figure 4.24. The work spindle has a screwed nose for attaching work holding devices such as a three-jaw chuck and catch plate. This spindle is also hollow and is provided with a taper at the front for accepting a centre when work is required to be held between centres. Rotation of the work spindle and therefore the work is obtained by turning the index crank.

Due to the ratio between the worm and the wormwheel, this means that if the crank is turned once then the work will turn through $\frac{1}{40}$ of a revolution. It follows then that when the crank is turned 40 times the work will rotate one complete revolution. So that sub-divisions of $\frac{1}{40}$ of a revolution can be obtained an index plate is provided which consists of a series of hole circles, each circle having a different number of holes. By locating the plunger (carried by the index crank) into one of these holes a proportion of $\frac{1}{40}$ of a revolution can be obtained in terms of so many holes in a specified hole circle. The index plates frequently used are of the Brown & Sharpe type each having the following hole circles:

Plate No. 1   15   16   17   18   19   20
Plate No. 2   21   23   27   29   31   33
Plate No. 3   37   39   41   43   47   49

Direct indexing   This is the easiest method of indexing and makes use of the direct index disc fitted on the front end of the work spindle. This disc has a number of slots (usually 24) around its periphery and may be locked into position by means of a plunger. When using this method the work is disengaged from the wormwheel and the work spindle is turned by hand through the required number of slots in the index disc. Needless to say, the divisions to be indexed must be an exact sub-multiple of the number of slots in the disc.

Simple indexing   Where direct indexing cannot be accommodated, simple indexing is used and probably represents the most commonly-used method. To index by this method the number of crank turns must first be determined. The reader will recall that one turn of the index crank will turn the work spindle, and hence the work, through $\frac{1}{40}$ of a revolution. If, now N equals the number of divisions to be indexed, then the number of crank turns will be N divided by $\frac{1}{40}$ which may be written as

$$\text{Crank turns} = \frac{40}{N}$$

Example: Determine the indexing for the following number of divisions, using Brown & Sharpe index plates:

(a) 7   (b) 28   (c) 82

(a) $\frac{40}{N} = \frac{40}{7} = 5\frac{5}{7} = 5\frac{15}{21}$

> Therefore, indexing = 5 crank turns plus 15 holes in a 21-hole circle.
>
> (b) $\dfrac{40}{N} = \dfrac{40}{28} = 1\dfrac{12}{28} = 1\dfrac{3}{7} = 1\dfrac{21}{49}$
>
> Therefore, indexing = 1 crank turn plus 21 holes in a 49-hole circle.
>
> (c) $\dfrac{40}{N} = \dfrac{40}{82} = \dfrac{20}{41}$
>
> Therefore, indexing = 20 holes in a 41-hole circle.

Use of sector arms

When a large number of indexings have to be carried out, the plunger movement around the hole circle may be controlled by the use of *adjustable sector arms*. These not only avoid the tedious counting of holes but also eliminate the risk of miscounting holes.

Consider the milling of seven slots around the periphery of a disc. As shown by (a) in the previous example this indexing will be five crank turns plus 15 holes in a 21-hole circle. After the first slot has been milled one sector arm is positioned against the plunger in the first position (Figure 4.25(a)), the other arm then

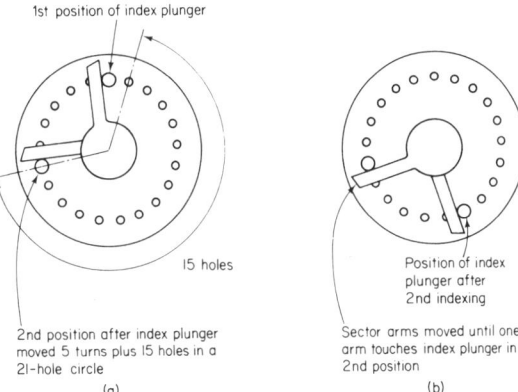

Figure 4.25 Use of sector arms

being adjusted to span a crank movement of 15 holes. It must be borne in mind that the actual number of holes between the sector arms will be 16, since the first hole is never counted. After five turns of the crank, the plunger is then located adjacent to the second sector arm, thus completing the first indexing. The sector arms, locked as a pair, are now moved clockwise until one arm touches the re-located plunger, as shown in Figure 4.25(b). When the second slot has been milled the previous plunger and sector movement is repeated to mill the remaining slots.

Angular indexing

The indexing of some components may be specified in terms of an angular movement rather than a number of divisions.

Since one turn of the index crank equals $\frac{1}{40}$ of a turn of the work, then angular movement of the work will be $\frac{1}{40} \times 360 = 9°$. Therefore the angular indexing becomes

$$\text{Crank turns} = \frac{\text{Angle to be indexed}}{9}$$

> *Example:* Determine the indexing for the following angles on a dividing head equipped with Brown & Sharpe index plates:
>
> (a) 23° (b) $15\frac{1}{3}°$ (c) 11° 15′
>
> (a) $\dfrac{23}{9} = 2\frac{5}{9} = 2\frac{10}{18}$
>
> Therefore, indexing = 2 crank turns plus 10 holes in an 18-hole circle.
>
> (b) $\dfrac{15\frac{1}{3}}{9} = \dfrac{46}{3} \times \dfrac{1}{9} = \dfrac{46}{27} = 1\frac{19}{27}$
>
> Therefore, indexing = 1 crank turn plus 19 holes in a 27-hole circle.
>
> (c) $11° 15′ = 11\frac{15}{60} = 11\frac{1}{4}°$
>
> $\dfrac{11\frac{1}{4}}{9} = \dfrac{45}{4} \times \dfrac{1}{9} = \dfrac{5}{4} = 1\frac{1}{4} = 1\frac{4}{16}$
>
> Therefore, indexing = 1 crank turn plus 4 holes in a 16-hole circle

Differential indexing This is used when the required indexing cannot be obtained by using the methods already discussed, due to the limited range of hole circles available. The technique consists basically of making a close approximation to the required indexing and then compensating for the resulting error by slightly rotating the index plate by means of suitable gearing. The direction in which the index plate rotates relative to the index crank is important and depends upon whether the approximation is larger or smaller than the required indexing.

The general arrangement of a dividing head set up for differential indexing is illustrated in Figure 4.26. To enable the

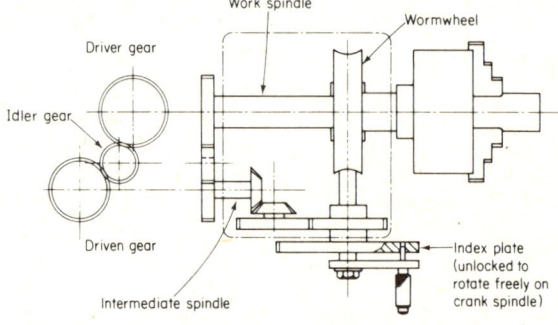

Figure 4.26 Dividing head set up for differential indexing

index plate to rotate it must be unlocked so that it is free to rotate on the crank spindle. The index plate also has fixed to it a gear that meshes with gearing coupled to the work spindle. Therefore, when the crank is turned with the plunger released the work spindle will rotate, thus causing a drive to be transmitted back to the index plate. The gearing between the work spindle and the intermediate spindle determines the extent to which the index plate rotates and must be determined depending on the difference between the approximate and true indexing fraction. The required direction of rotation of the index plate is obtained by using idler gears between the driver and driven gears.

> *Example:* Determine the differential indexing of 125 divisions using a dividing head equipped with Brown & Sharpe index plates. The following gears are available:
>
> 24 (two gears) 28 32 40 44 48 56 64 72 86 100 teeth
>
> Crank turns $= \dfrac{40}{N} = \dfrac{40}{125}$
>
> Approximate indexing $= \dfrac{40}{120} = \dfrac{4}{12} = \dfrac{9}{27}$
>
> Therefore, indexing = 9 holes in a 27-hole circle.
>
> If now $\dfrac{9}{27}$ is indexed 125 times, then
>
> $\dfrac{9}{27} \times 125 = 41\tfrac{2}{3}$ crank turns
>
> Since the index crank must make 40 turns for one revolution of the work, this means that when nine holes are indexed in a 27-hole circle the crank will move $1\tfrac{2}{3}$ of a turn too far. In order to correct this error the work spindle must now be geared to the index plate and arranged so that the index plate rotates 'opposite' to that of the index crank.
>
> Therefore, $\dfrac{\text{Drivers}}{\text{Driven}} = 1\tfrac{2}{3} = \dfrac{5}{3} = \dfrac{40}{24}$
>
> The 40-tooth gear will be fitted to the work spindle while the 24-tooth gear will be fitted to the intermediate spindle.

The rotary table This is an auxiliary table fitted to the milling machine table to enable circular forms to be machined. A typical rotary table is illustrated in Figure 4.27.

The rotary motion is obtained through a wormwheel and worm drive in similar fashion to that found on the dividing head. The periphery of the table is usually calibrated in degrees while sub-divisions of a degree are divided on a calibrated drum carried by the handwheel. In addition to this calibrated drum some rotary tables are fitted with an index plate and plunger so that they may be used as a dividing head. This feature is particularly useful for indexing work which might otherwise be difficult to hold using a conventional dividing head.

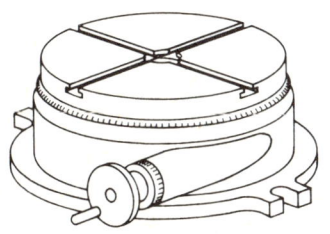

Figure 4.27 Rotary table

HELICAL MILLING

Helical features, such as the flutes found on twist drills, reamers and milling cutters, may be machined on a universal milling machine using a dividing head as part of the set-up. To cut a helix the work must rotate simultaneously with the longitudinal table feed and this is achieved by gearing the dividing head to the machine table, as shown in Figure 4.28. The gearing selected is dependent on the lead of the helix to be cut, the lead being the distance moved by the work while it makes one complete revolution. Before this gearing can be set up the 'lead of the machine' must be known. This is the lead produced when the gear ratio between the machine leadscrew and dividing head is 1:1. Therefore if a 40-to-1 dividing head is used then the lead of the machine will be

Lead of machine = 40 × pitch of table leadscrew

To determine the gear required to cut a specified helix the following relationship is used:

$$\frac{\text{Drivers}}{\text{Driven}} = \frac{\text{Lead of machine}}{\text{Lead to be cut}}$$

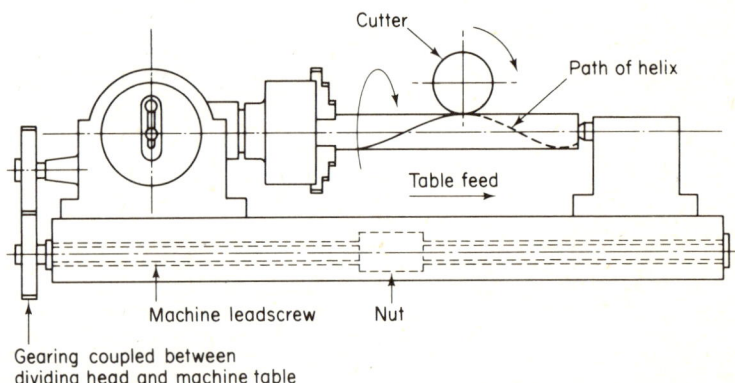

Figure 4.28 Helical milling

When a helix is cut on a horizontal milling machine the table must be set over to the helix angle so that the cutter lies along the helix, as shown in Figure 4.29(a). If this adjustment were not made then a considerable amount of cutter interference would be produced on the workpiece, resulting in a slot or

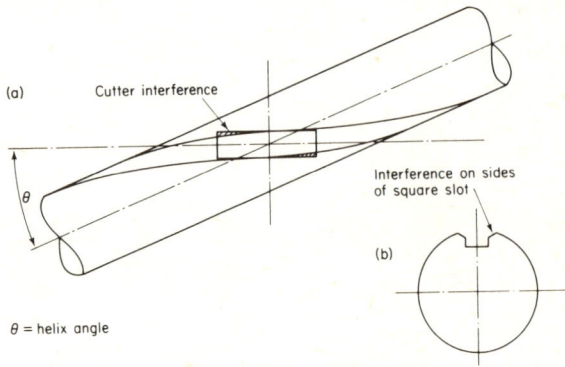

Figure 4.29 Effect of cutter interference

groove having a shape different from that of the cutter (Figure 4.29(b)). Even with the table swung through the helix angle some interference will occur, and this is more noticeable on slots having a square or rectangular cross-section. This is because to machine this type of slot on a horizontal milling machine it is usual to use either a side-and-face or a slotting cutter which are designed for cutting slots having straight parallel sides, while a helical slot has curved sides.

For a helical slot of given lead, the helix formed at the top of the slot will have a larger helix angle than the helix at the bottom of the slot. For this reason the machine table is set to the helix angle based on the mean diameter of the workpiece and is calculated from the following expression:

$$\tan \theta = \frac{\pi D}{\text{lead}}$$

where D = mean diameter of work, and θ = helix angle

This expression is derived from the development of a helix which is in the form of a right angle triangle as shown in Figure 4.30.

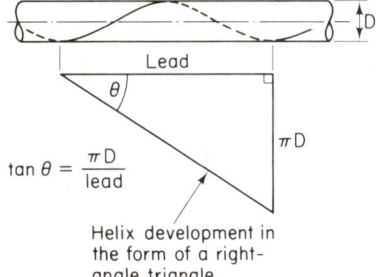

Figure 4.30 Helix development

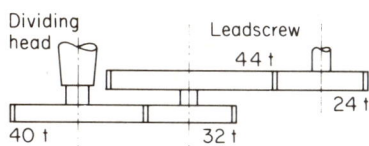

Figure 4.31 Compound gear train

Example: A helical slot is to be cut on a universal milling machine. The slot is specified as being 6 mm deep on a 50 mm-diameter steel bar and is to be cut with a 550 mm lead. If the machine table leadscrew has a 6 mm pitch, determine the gears required and the necessary machine table adjustment. Assume a set of Brown & Sharpe gears is available.

Lead of machine = 40 × pitch of table leadscrew
= 40 × 6
= **240 mm**

Lead to be cut = 550 mm

$$\frac{\text{Drivers}}{\text{Driven}} = \frac{\text{Lead of machine}}{\text{Lead to be cut}} = \frac{240}{550} = \frac{24}{55}$$

$$= \frac{6}{11} \times \frac{4}{5} = \frac{24}{44} \times \frac{32}{40}$$

This results in a compound gear train which will be set up as shown in Figure 4.31.

Mean diameter (D) of the workpiece = 50 − 6 mm
= 44 mm

Therefore, $\tan \theta = \dfrac{\pi D}{\text{Lead}} = \dfrac{\pi \times 44}{550} = \dfrac{138.23}{550}$

= 0.2513

$\theta =$ **14° 6′**

GRINDING Grinding is basically a finishing process, designed to provide close dimensional accuracy with improved surface finish, usually on work that has been previously machined. Material is removed by means of a grinding wheel which may be regarded as a multi-toothed cutter. These cutting teeth or edges are in the form of a large number of abrasive particles which are securely held together throughout the wheel by means of a suitable bond.

When in operation the behaviour of a grinding wheel is different from that of other multi-toothed cutters, e.g. milling cutters. A grinding wheel is so designed that as soon as the abrasive particles become blunted they break away from the bond to reveal fresh sharp abrasives. Thus a grinding wheel has a self-sharpening action. It should be appreciated, however, that this will happen only if the correct type of wheel is chosen for the material to be ground.

GRINDING WHEEL SPECIFICATION

When specifying a grinding wheel for a particular operation the following constructional features of the wheel must be considered:

(i) Abrasive material
(ii) Abrasive size
(iii) Grade of the bond
(iv) Structure
(v) Type of bond.

Abrasives

Basically two types of abrasive are used, these being *aluminium oxide* and *silicon carbide*. Aluminium oxide is softer and tougher compared with silicon carbide and this makes it suitable for grinding materials of high tensile strength. On the other hand, silicon carbide is much harder and brittle, thus making it suitable for materials of low tensile strength, such as cast iron, brass and aluminium alloys.

Sizes of abrasive

The size of the abrasive determines the finish produced on the work, the smaller the abrasive the finer the finish. The size of the abrasive particles, sometimes known as *grits*, is designated by a number which is the number of holes per 25 mm of mesh through which the grits will pass. These numbers, which range from 8 to 600, are further grouped to classify the grit size as being coarse, medium, fine, or very fine, as indicated below:

Coarse	8	10	12	14	16	20	24
Medium	30	36	46	60	80	100	120	150
Fine	180	220	240	280
Very fine	320	400	500	600

Grade

The term *grade* is used to denote the ability of the bond to retain the abrasive. The ideal bond is one which allows the abrasive particles to break away as soon as they become dull, thereby revealing fresh sharp abrasive particles. When this breaking down occurs easily the wheel is said to be 'soft' and when the abrasives can be broken away only with considerable force the wheel is said to be 'hard'. The grade of a grinding wheel is specified according to a system of capital letters, as follows:

Soft	E	F	G	H	I
Medium	J	K	L	M
Hard	N	O	P	Q	R
Very hard	S	T	U	V	W	X	Y	Z

The grade of a wheel is chosen according to the material to be ground and in general it may be said that the harder the material the softer the wheel. This is so because when grinding hard materials the abrasive particles become dulled quickly which means that the abrasives must in their turn be broken away quickly.

Structure The *structure* of a grinding wheel refers to the spacing of the abrasives within the bond. When the abrasives are spaced widely apart, the wheel is said to have an 'open' structure, while close spacing of the abrasives is referred to as a 'dense' structure. The significance of the structure is mainly to provide clearance for the chips between the abrasives, which is important in rough grinding operations where rapid stock removal is required. Furthermore, with an open structure wheel the heat generated will be less since there are less grits in contact with the work compared with a wheel having a dense structure. The structure of a grinding wheel is specified by a numbering system as shown below:

DENSE 2 3 4 5 6 7 8 9 10 11 12 13 14 15 16 OPEN

Bond As already stated, the function of the bond is to hold the abrasive particles together. In general four types of bond are used, these being

(i) Vitrified (ii) Resinoid (iii) Rubber (iv) Shellac.

The majority of grinding wheels have a *vitrified* bond, and these are manufactured from clay and fusible materials. After mixing with the abrasive the wheel is then pressed to shape in moulds, followed by drying to remove the moisture. After drying, the wheel is then finished by being baked in a kiln.

The *resinoid* bond is used for wheels required to run at high speeds, therefore making them suitable for rapid stock removal. Furthermore, the resinoid bond is used where wheels are subjected to shock loading as found, for example, in fettling operations.

Rubber used as a bonding material produces a very strong wheel. The main application of this bond is for very thin cutting-off wheels which may have to withstand small side deflections.

The *shellac* bond may be used as an alternative to the rubber bond, but its main application is for work requiring a fine surface finish.

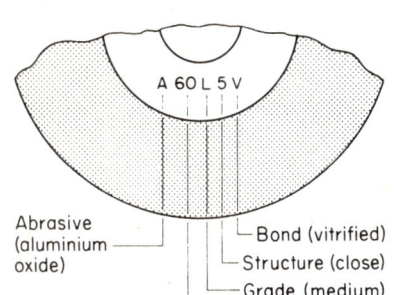

Figure 4.32 Grinding wheel marking system

Marking system The grinding wheel specification is always indicated on the side of the wheel by using the letter and numbering system already described. Figure 4.32 illustrates how a grinding wheel is specified using this system.

GRINDING WHEEL SHAPES

Manufacturers of grinding wheels provide wheels having many shapes and sizes for a wide range of applications. Some of the more common shapes are illustrated in Figure 4.33.

The *straight wheel*, the most widely used type, is found on 'off-hand' surface and cylindrical grinding machines. A modification to this type is the *tapered wheel*, the sides of which taper towards the periphery where the consequent reduction of

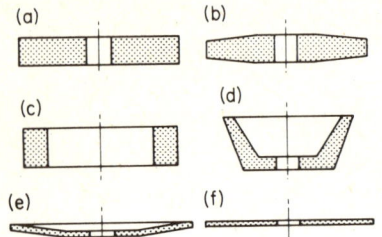

Figure 4.33 Grinding wheel shapes, (a) straight, (b) tapered, (c) cylinder, (d) flaring cup, (e) dish, (f) cutting off

its mass allows this wheel to be run at a higher speed than the straight type. The *cylinder wheel* is used on vertical spindle surface grinding machines, cutting taking place on the side rather than at the periphery. This type of wheel is generally used where large surface areas have to be ground. The *flaring cup wheel* and the *dish wheel* are normally used for tool and cutter grinding, the dish wheel being particularly useful for grinding the front face of form relieved cutters and the flutes of taps. The *cutting-off wheel*, sometimes known as an *elastic wheel*, is used for grinding narrow slots and cutting off hardened stock.

GRINDING WHEEL SPEEDS

For specific grinding processes the peripheral speed of the wheel will be specified by the manufacturer. Typical examples are shown in the following table:

Process	Wheel speed m/min
Surface grinding	1500–1800
Cylindrical grinding	1650–2000
Internal grinding	1650–2000
Tool grinding	1700

The peripheral speed of the wheel is constant irrespective of its diameter. Since the machine spindle speed is measured in revolutions per minute (rev/min) this means that the wheel speed in rev/min must be known, and may be obtained from the following expression:

$$N = \frac{1000S}{\pi D}$$ where S = peripheral speed of wheel in m/min

d = diameter of grinding wheel in mm.

The 'maximum' speed (rev/min) is always indicated on wheels over 55 mm in diameter, and in the interests of safety it is important that this maximum speed is never exceeded. Furthermore, the reader should appreciate that if a wheel is operated at a speed other than its recommended speed, then this will have the effect of altering the grade of the wheel. For example, if a hard wheel is operated slowly it will behave as if it is soft.

GRINDING WHEEL MOUNTING

Before mounting, grinding wheels should always be inspected for any defects and cracks. This is important since stresses set up within wheels due to the high speed of rotation have caused them to burst, resulting in serious injury to grinding machine operators.

Basically, two methods are used to mount grinding wheels, the choice depending on the use to which they are put.

For 'off-hand' or general purpose tool grinders the method shown in Figure 4.34(a) is used. The wheel is secured between two flanges, one flange being keyed to the machine spindle so that a positive drive is transmitted. It is essential that these flanges are relieved so that clamping is effected only on the outer rim, and also that blotting paper discs are interposed between the flanges and the sides of the wheel. The purpose of these discs is to ensure that the wheel is not subjected to any unnecessary clamping stresses which might otherwise occur if the flanges made direct contact with the sides of the wheel. Grinding wheels should never be forced onto their spindles, if difficulty is experienced then the lead bush may be gently scraped to produce a sliding fit.

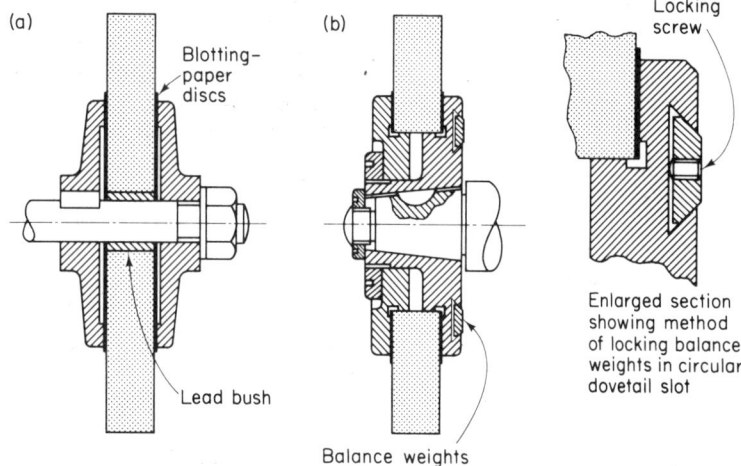

Figure 4.34 Mounting of grinding wheels

For precision grinding operations, such as those found on surface and cylindrical grinding machines, the wheel mounting method shown in Figure 4.34(b) is used. Here the wheel is mounted on a special adaptor or collet, the assembly being keyed to the tapered spindle of the machine. Sliding weights are provided on one side of the collet to facilitate wheel balancing.

BALANCING AND TRUEING OF GRINDING WHEELS

Grinding wheels are always initially balanced by the manufacturer to a sufficient accuracy for wheels to be fitted to 'off-hand' grinding machines. However, wheels for precision grinding operations require a finer degree of balance since any slight out-of-balance may impair the quality of the workpiece. After having been mounted on their collet such wheels should first be trued. Trueing is an operation carried out to render the wheel truly cylindrical and is achieved by passing a diamond across the face of the wheel while it rotates at normal operating

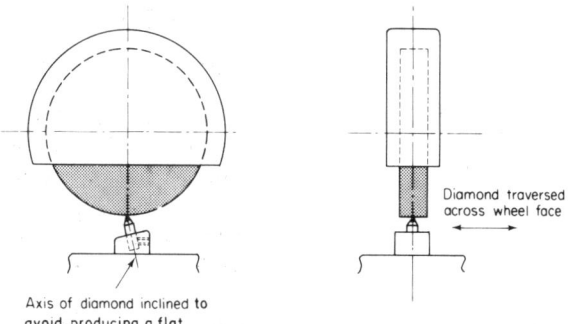

Figure 4.35 Trueing a grinding wheel (surface grinding)

speed. As well as trueing the face of the wheel, it may also be necessary to true the sides to render them parallel and at 90° to the axis of rotation. Wheel trueing is illustrated in Figure 4.35.

After trueing, the wheel is removed from the machine and fitted to a balancing mandrel, and the assembly is placed onto a balancing unit consisting basically of a pair of levelled knife edges, as shown in Figure 4.36. If the wheel is free to roll, it will always come to rest with the heavy side downwards, which means that the balance weights must be moved towards the light side. These weights are continually adjusted until the wheel remains static in any position. After balancing, the wheel is re-fitted to the machine and trued once more before being put into operation.

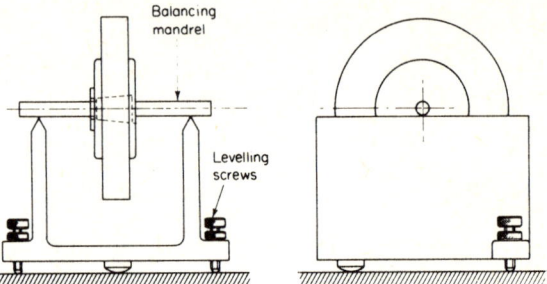

Figure 4.36 Grinding wheel balancing

GRINDING WHEEL DRESSING

After prolonged use grinding wheels may lose their cutting efficiency due to particles of swarf and broken dulled grains becoming embedded in the wheel face. Dressing is a technique used to clear these particles and it may be carried out in a variety of ways depending on the application of the wheel. Surface and cylindrical grinding machine wheels are dressed by using a diamond in exactly the same way as for 'trueing'.

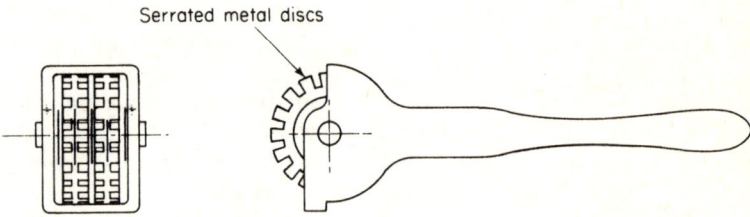

Figure 4.37 Star wheel-dresser

With 'off-hand' grinders, where the workpiece surface finish is not of prime importance, a *star wheel-dresser* is used. This consists of a series of serrated metal discs mounted on a handle, as shown in Figure 4.37. In use this dresser is pushed against the rotating grinding wheel, thus causing the discs to rotate and in so doing they will remove any trapped swarf or abrasive particles. As an alternative to the star wheel-dresser an abrasive stick may be used which has a composition similar to that of the grinding wheel.

WHEEL FORMING

Many precision grinding operations involve grinding a profile onto the workpiece, which means that in the majority of cases the same profile must be reproduced onto the periphery of the wheel.

Crush forming

With this method a hardened steel roller, having the same profile on its periphery as that to be ground, is forced into the grinding wheel while it rotates at a relatively low speed (about 180 rev/min). Due to the compressive forces set up, the abrasive grains are broken away until the full depth of the profile is obtained. It is important that a copious supply of cutting fluid is always directed between the crushing roll and the wheel. The principle of the process is shown in Figure 4.38.

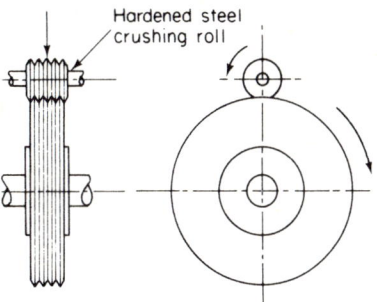

Figure 4.38 Crush-forming imparting a thread profile onto a grinding wheel

Forming using a pantograph

A *pantograph* consists of a system of linkages, as shown in Figure 4.39(a), designed to reproduce a profile onto the grinding wheel at a diminished magnification from an enlarged template. A property of the pantograph is that no matter what configuration the linkages take up, a straight line will always pass through the stylus, the fixed point P and the dressing diamond, as shown in Figure 4.39(b). One advantage of the pantograph is that although the template is made to a high degree of accuracy, any errors that are present will be correspondingly decreased on the wheel, depending on the scale reduction used.

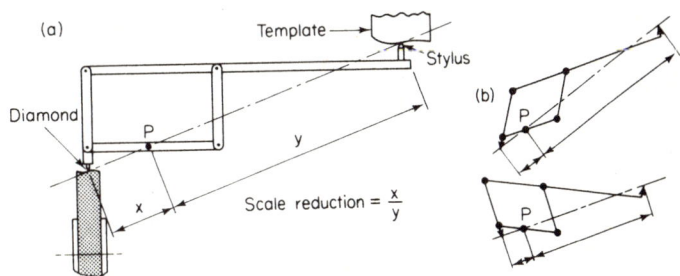

Figure 4.39 Wheel-forming using a pantograph

CHARACTERISTICS OF GRINDING

Due to the nature of the grinding process there are certain characteristics which are unlike those found in other multi-tooth cutting processes.

Arc and area of contact During surface and cylindrical grinding operations the *arc of contact* between the wheel and the work is usually quite small, as shown in Figure 4.40(a)(b). For example, when surface grinding using a 200 mm-diameter wheel with a 0.01 mm depth of cut the length of the arc will be about 2 mm, resulting in a relatively small area of contact. Under these conditions the wheel should have a small abrasive size and a hard bond.

Grinding operations using cylinder wheels produce a large *area of contact* (Figure 4.40(c)). These conditions require the wheel to have a large abrasive size together with a soft bond. The soft bond is required in this case since the abrasives will become dulled more quickly owing to the greater amount of work that they have to do over a large area. Furthermore, by using a large size of abrasive the heat dissipation will be improved.

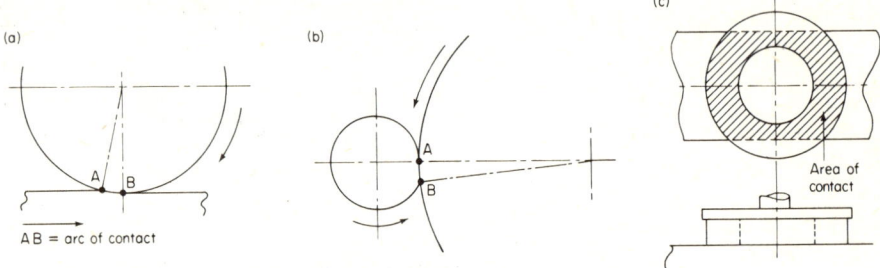

Figure 4.40 Arc and area of contact, (a) surface grinding, (b) cylindrical grinding, (c) surface grinding using a cylinder wheel

When grinding (surface or cylindrical) up to a face or shoulder (Figure 4.41) the side of the wheel should be relieved so that the area of contact is reduced.

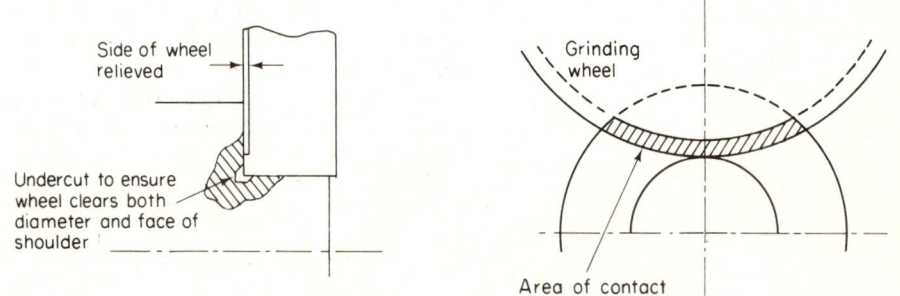

Figure 4.41 Grinding up to a shoulder in cylindrical grinding

Work speeds and feeds Work speeds for surface grinding are usually between 20 and 30 m/min. The exact speed chosen will depend mainly on the depth of cut, finish required and the material to be ground. If too slow a speed is used then overheating of the work may occur, while too high a speed may cause vibration. If overheating is a problem then a softer wheel with a larger abrasive size should be used. The cross-feed is generally quite small, in the order of about 3 mm per pass (peripheral surface

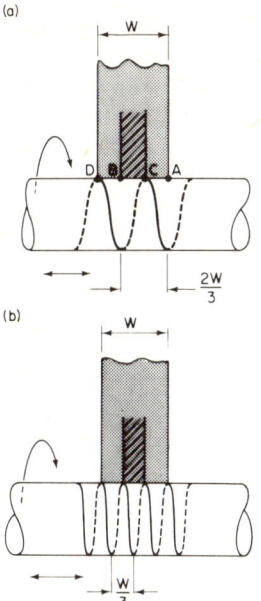

Figure 4.42 Work feed for cylindrical grinding

grinding), although this figure may be increased for roughing operations.

The work peripheral speed for cylindrical grinding is generally about 24 m/min. Since the area of contact is smaller, compared with surface grinding, the problems of overheating are not so apparent. The rate of feed for cylindrical grinding deserves attention since this will affect the way in which the wheel wears. For roughing operations a feed rate of approximately two-thirds the width of the wheel per revolution of the work is preferred. Under these conditions the wheel will tend to wear concave. This will be readily understood by making reference to Figure 4.42(a).

When the work traverses to the right, most of the wheel wear will be concentrated along the length DB. On the return pass, when the work traverses to the left, most of the wear now occurs along AC. This means that the centre part BC will be subjected to twice as much wear as the outer parts resulting in the wheel wearing slightly concave. If the feed rate is approximately one-third the width of the wheel per revolution (Figure 4.42(b)), then the wheel tends to wear convex. For finishing operations a feed rate of not less than one-half the width of the wheel is used.

GRINDING FLUIDS

It is common with many precision grinding operations that an adequate supply of fluid is supplied between the wheel and work. Ideally a flattened supply nozzle which directs the fluid across the whole width of the wheel should be used.

For maximum cooling, water is the most efficient, but it has obvious disadvantages in that the machine parts are vulnerable to corrosion. However, grinding fluids with a water base are available and these incorporate a high percentage of rust inhibitors. Soluble oils are frequently used having a dilution of about one part oil to 50/60 parts water. Soluble oils for heavy-duty grinding operations are also available with 'extreme pressure' properties due to the inclusion of sulphur and chlorine additives.

GRINDING FAULTS

Due to the incorrect choice of wheel or incorrect grinding conditions the following two faults may occur.

Loading

This fault occurs when the spaces between the abrasive grains become clogged with particles of the metal being ground. The resulting effect is that the grains do not project sufficiently to promote efficient cutting. Loading is usually associated with the grinding of soft metals, especially if an open-structured wheel is used.

Glazing

This condition, easily recognised by the shiny appearance on the face of the wheel, occurs when the abrasive grains become dulled and do not break away from the bond to reveal fresh grains. Wheels prone to glazing are therefore too hard for the material being ground. Glazing may be reduced by increasing the wheel or work speed which tends to make the wheel behave as though it is soft.

CUTTER GRINDING

Tools such as lathe tools and drills are generally sharpened on off-hand grinding machines. Milling cutters, reamers and taps however require a special grinding machine known as a *tool-and-cutter grinder* (Figure 4.43). To a certain extent this machine is similar in construction to that of a surface grinder, the main difference being that the wheel head may be swivelled

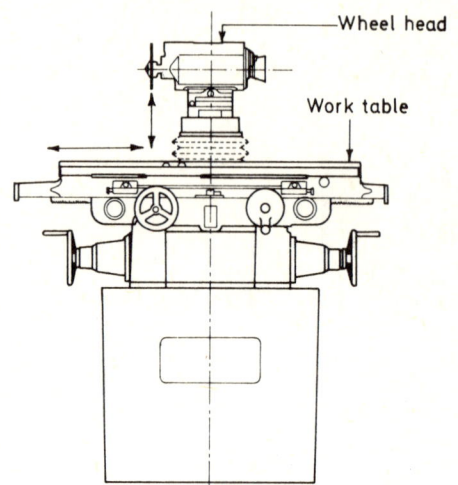

Figure 4.43 Tool-and-cutter grinder (Elliot Machine Tool Group)

in the vertical and horizontal planes as well as being able to be raised up or down. The table on these machines is supported on hardened steel balls which results in the table having a floating action thus giving the operator a light sensitive control. Furthermore a wide range of attachments is available to accommodate the many different tool angles that have to be ground.

Cutter setting

When *horizontal milling cutters* are being ground, they are always mounted on a mandrel held between centres. The tooth to be ground is supported by a spring-loaded index finger, so that after each tooth has been ground the cutter is rotated to index the next tooth.

To grind the primary clearance a cup wheel is normally used which may be set relative to the cutter in one of two ways, as shown in Figure 4.44. The method shown in Figure 4.44(a) is

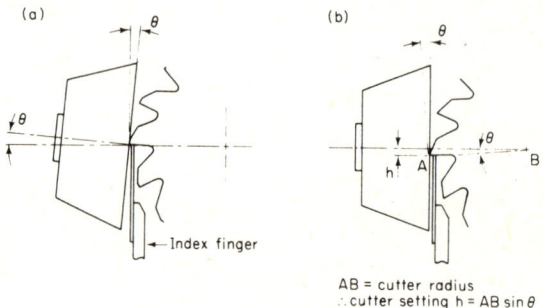

AB = cutter radius
∴ cutter setting $h = AB \sin \theta$

θ = primary clearance angle

Figure 4.44 Grinding the primary clearance on cutters used for horizontal milling

the easier since no calculation is required and also more space is provided for the index finger since the wheel is swung away from the tooth.

To grind the flutes on a *helical cutter*, e.g. a slab mill, a disc wheel is normally used. This time the cutter is held against the finger and the table traverse operated so that the cutter follows the finger. Care must be exercised since two movements must be controlled simultaneously. The correct primary clearance is established by setting the cutter tooth below the centre of the grinding wheel, as shown in Figure 4.45.

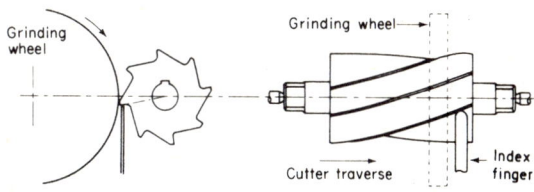

Figure 4.45 Grinding the flutes of a helical slab mill

When grinding *form-relieved cutters*, these must be ground on the front face to preserve their form. This means that an index finger cannot be used since the previously-ground teeth will interfere with the indexing, as shown in Figure 4.46(a). Instead, the cutter is indexed from a separate index disc fitted to the same mandrel as the cutter (Figure 4.46(b)).

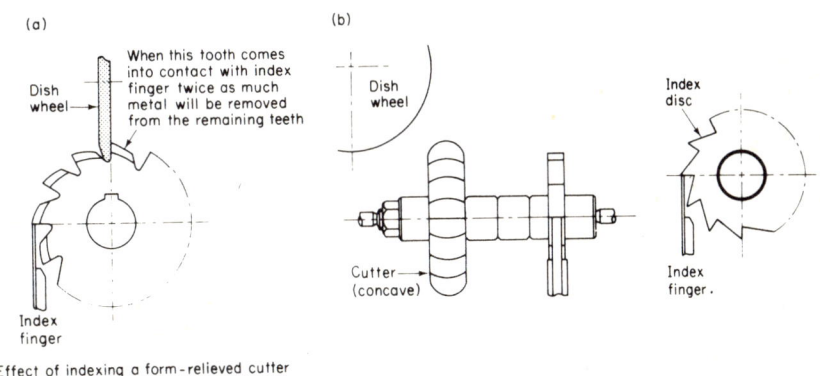

Figure 4.46 Grinding form-relieved cutters

The settings for *end-mills* and *shell-mills* are carried out in the same way, the only difference being that these cutters are held in chucks/collets fitted to a workhead.

GRINDING SAFETY

Grinding operations have always been regarded as a source of potential danger. The following represents a summary of certain precautions that should be observed. For a more detailed account the reader is advised to consult the Health and Safety at Work publication No. 4, *Safety in the Use of Abrasive Wheels* (published by HMSO) from which the following has been extracted with permission:

(1) Mounting and balancing of grinding wheels should be carried out only by competent and trained persons.
(2) The maximum operating speed marked on the wheel should in no circumstances be exceeded.
(3) When starting a new wheel it should be run at normal operating speed for a short period with operators and others standing clear during the trial run.
(4) Grinding wheels should always be fitted with a guard which should be positioned and properly adjusted before the wheel is run.
(5) Work rests on off-hand grinders should be adjusted as close as possible to the wheel.
(6) Eye protectors should be worn in all dry-grinding operations, or alternatively, transparent screens should be fitted to the machine to intercept sparks and particles.

SUMMARY

During metal cutting with single-point tools three forces act at the tip of the tool. These forces may be measured using a tool-force dynamometer. The vertical cutting force is the largest and may be used to determine the power consumed in cutting. Different constructions are used for single-point tools. Tungsten carbide tools may be used in the form of a brazed insert or in the form of throw-away tips. On production machining quick-change and pre-set tooling is often used.

Cutters used for milling are classified as being of the fluted, form-relieved or inserted-tooth types. Cutters may be operated using either the up-cut or down-cut (climb) milling methods. Parallel surfaces may be machined simultaneously by straddle milling, while profiles may be machined by gang milling. The dividing head is used for indexing and may also be used for machining helical slots. The rotary table is used for machining circular forms.

Grinding is a multi-tooth cutting process. A grinding wheel is classified according to its abrasive, grade, structure and bond. Grinding wheels must be properly mounted and balanced. Profiles may be imparted onto grinding wheels either by crush forming or by using a pantograph. Incorrect grinding conditions may produce faults in the wheel known as 'loading' and 'glazing'. Tool and cutter grinders are used for sharpening tools such as milling cutters, reamers and taps. Strict safety precautions must be observed when using grinding wheels.

QUESTIONS

(1) With the aid of a diagram show the three principal forces that act at the tip of a single-point tool.

(2) During a test using a lathe tool dynamometer the following results were recorded:

Spindle speed 500 rev/min   Depth of cut 5 mm
Blank dia. of work 60 mm   Cutting force 1650 N

Determine the power consumed at the tool point.

(3) The following information relates to the machining of a certain steel bar in a centre lathe:

Cutting speed 30 m/min
Depth of cut 4 mm
Feed 0.2 mm/rev

Determine the metal removal rate in mm^3/min.

(4) (a) With the aid of a sketch describe a typical design for a throw-away tip toolholder.
(b) State the advantages of using throw-away tip tooling.

(5) A high-speed steel tool was found to have an average life of one hour between re-grinds when used in a lathe for roughing cuts on steel bar at 30 m/min. Given that $n = 0.12$ for roughing and 0.10 for finishing, calculate the probable life for the tool between re-grinds when used for finishing cuts on the same steel bar.

(6) Using suitable sketches show clearly the difference between

(i) Up-cut milling (ii) Down-cut milling.

(7) By means of sketches show a typical application for each of the following types of milling cutter:

(i) Side-and-face cutter (iii) Slot drill
(ii) End mill (iv) Slab mill.

(8) With reference to milling cutters, what is meant by the term 'form relieved cutter'.

(9) A 100 mm-diameter side and face cutter having 12 teeth is used to machine a slot in an aluminium alloy at 45 m/min. If the feed per tooth is 0.08 mm, determine the machine table feed in mm/min.

(10) Describe, using a sketch, the difference between

(i) Straddle milling (ii) Gang milling.

(11) Using Brown & Sharpe index plates, determine the indexing for the following numbers of divisions:

(i) 9 (ii) 34 (iii) 96 (iv) 108.

(12) Explain what is meant by the following terms with reference to grinding wheels:

(i) Grade (ii) Bond (iii) Structure.

(13) Show by means of a clear diagram the correct method of mounting a grinding wheel.

(14) Explain the importance of grinding-wheel balancing.

(15) Explain why grinding wheels have to be 'dressed' periodically.

(16) Describe with the aid of a diagram how a profile may be imparted to a grinding wheel by 'crushing'.

(17) With reference to grinding wheels, what is meant by the terms

(i) Loading (ii) Glazing.

(18) By means of suitable diagrams indicate how the following cutters are sharpened:

(i) Slotting cutter (ii) Form-relieved cutter.

5 Capstan and turret lathes

INTRODUCTION

Although the *centre lathe* is an extremely versatile machine tool, it is however not suitable for the economic manufacture of parts on a repetitive basis. This is due to the fact that centre lathes require skilled operators and that a large proportion of the total machining time is taken up with tool changing. *Capstan* and *turret lathes* are therefore a development of the centre lathe and are designed to manufacture turned parts rapidly using pre-set tooling. Furthermore, once set by a toolsetter these machines may be operated by semi-skilled operators.

The purpose of this chapter is to consider the construction of capstan and turret lathes together with their special tooling arrangements. Also to introduce the reader to the principles of process planning associated with these machines.

GENERAL MACHINE DESIGN

The general design of capstan and turret lathes is similar to the centre lathe except that the tailstock is replaced by a turret. This turret is fitted with a series of pre-set tools and can be indexed to present different tools to the workpiece. In addition the cross-slide has five tool stations, a front four-way tool post and single rear tool post. The movement of these tools is controlled by adjustable stops.

The general arrangement for both capstan and turret lathes is illustrated in Figure 5.1.

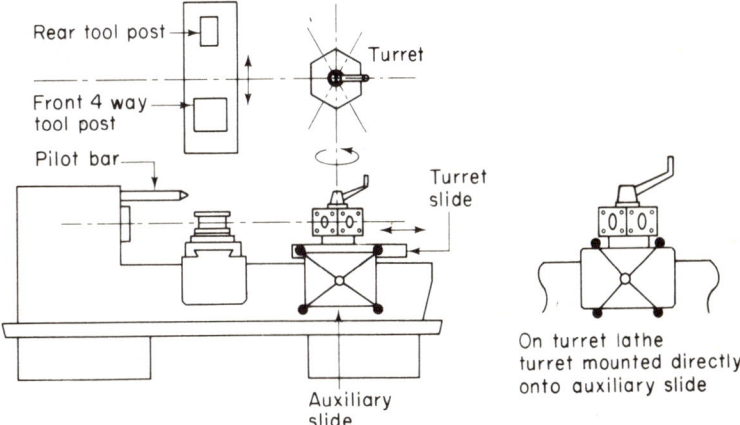

Figure 5.1 Capstan and turret lathes

On the capstan lathe the turret is carried on a separate slide which, when moved back (operated by a star wheel), will index the turret to the next tool station. This slide, together with the turret, is attached to an auxiliary slide which is locked to the bed after the machine has been set up. On the turret lathe, the turret is mounted directly onto the auxiliary slide. This creates a far more rigid construction, thus making turret lathes more suitable for heavy work, e.g. the machining of castings and forgings.

The headstock transmission on capstan and turret lathes differs slightly compared with that of the centre lathe. With the former machines it is important that speed changing is affected in the shortest possible time. This is usually achieved by using a constant-mesh, all-geared headstock, where change of speed is performed by operating multi-plate friction clutches.

WORK HOLDING DEVICES

In general these are similar to those found on the centre lathe, with the exception that more emphasis is put on speed of work loading and ease of operation.

Collet chuck

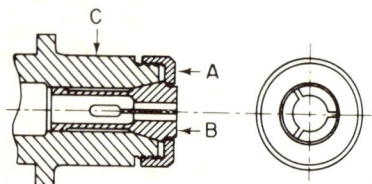

Figure 5.2 Principle of collet chuck operation

Much of the work carried out on capstan lathes is machined from bar which is fed through the hollow spindle. For this type of work collet chucks are generally used. The principle of the collet chuck is shown in Figure 5.2. When the cap A is turned this will cause the collet B to push against the adaptor C thus causing the collet to close down and grip the work.

Collets should only be used to hold workpiece sizes for which they were designed. However, collets are available for use with inserts so that a range of work sizes can be accommodated by using a master collet, as shown in Figure 5.3. Furthermore, collets are also available for holding square and hexagonal bar.

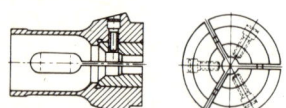

Figure 5.3 Master collet and inserts (H. W. Ward & Co)

The majority of collet chucks are hand-operated, but in some cases pneumatic operation is preferred. This method of operation is often adopted to reduce operator fatigue which might otherwise occur when opening and closing chucks manually.

Jaw chucks

The most common chuck used is the three-jaw self-centring type. When this chuck is used on capstan and turret lathes, the jaws are in two parts, a base jaw that always remains in the chuck body, and a removable gripping jaw. This design enables the hardened gripping jaw to be replaced by soft jaws. These are unhardened steel blanks specially designed to hold irregularly-shaped components, and they are fitted to the base jaws of the chuck, as shown in Figure 5.4(a). For certain classes of work two-jaw self-centring chucks are used, a typical application being shown in Figure 5.4(b).

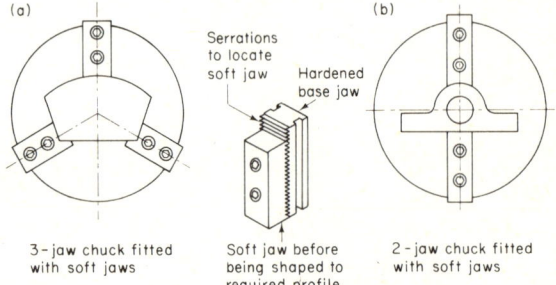

Figure 5.4 Jaw chucks

Large chucks used on capstan and turret lathes are often power-operated by electric, pneumatic or hydraulic methods which greatly reduce operator fatigue.

STANDARD TOOLING Manufacturers of capstan and turret lathes provide a wide range of tooling that can be pre-set for specific machining operations. The complete range of tools available is extremely large, therefore only some of the more widely used types are dealt with in this section.

Bar stops Bar work on capstan lathes is always set to a stop before machining commences so that the bar is in the correct position relative to the cutting tools. These stops usually consist of the adjustable type (Figure 5.5(a)). Where a hole has to be drilled, the stop may be combined with a centre drilling toolholder (Figure 5.5(b)). Here the stop is mounted on a swinging plate which, when raised, permits the centre drill to be brought into operation by a rack-and-pinion mechanism.

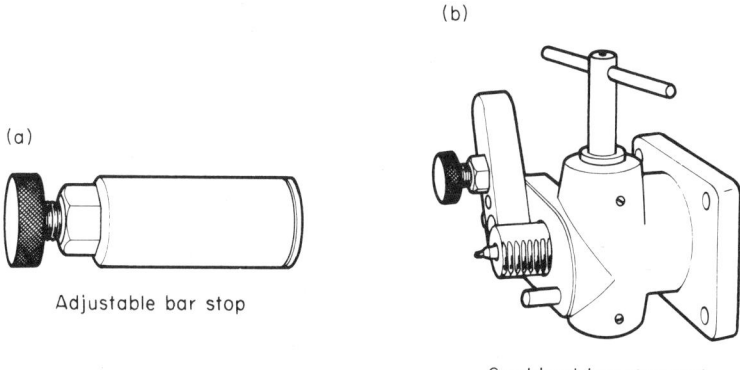

Figure 5.5 Bar stops (H. W. Ward & Co)

Toolholders for parellel turning These types of toolholders are many in number. The *roller steady-turning toolholder* (Figure 5.6(a)) is used from the turret

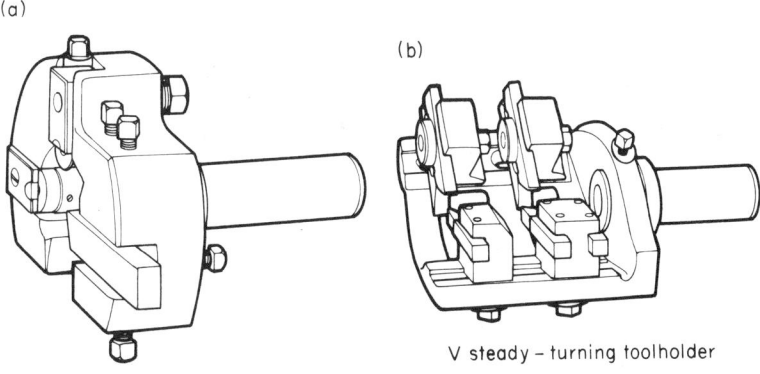

Figure 5.6 Roller and V steady-turning toolholders (H. W. Ward & Co)

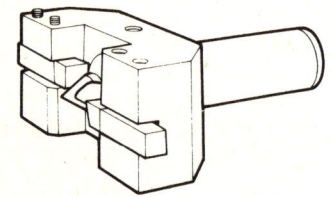

Figure 5.7 Start-turning toolholder (H. W. Ward & Co)

for machining small diameter work. So as to avoid deflection of such workpieces, support rollers are provided which may either lead or trail the cutting tool. The rollers are used leading the cutting tool when the work has been previously machined, so that concentricity is maintained. When the roller trails the cutting tool this often brings about an improvement in surface finish due to the burnishing action between the roller and machined surface. As an alternative to roller steady support, fixed V steadies may be used, which are generally employed for machining brass and other free-cutting materials. The *V steady-turning toolholder* (Figure 5.6(b)) may be used for machining two diameters simultaneously.

The *start-turning toolholder* (Figure 5.7) is used for turning short lengths to prepare the work for use with a roller steady-turning toolholder. A centre is used to support the work which slides into a spring-loaded shank as the turret moves forward. Two cutting tools are used diametrically opposite so that the cutting forces are neutralised.

The *knee-turning toolholder* (Figure 5.8(a)) is operated from the turret and is generally used for large-diameter workpieces which are outside the range of other forms of tooling. Due to the overhang on these toolholders, they are fitted with a guide bush which engages with the pilot bar projecting from the headstock. This therefore makes for a much more rigid set-up. Knee-turning toolholders are also available whereby a combination of tools including boring bars may be used simultaneously, as shown in Figure 5.8(b).

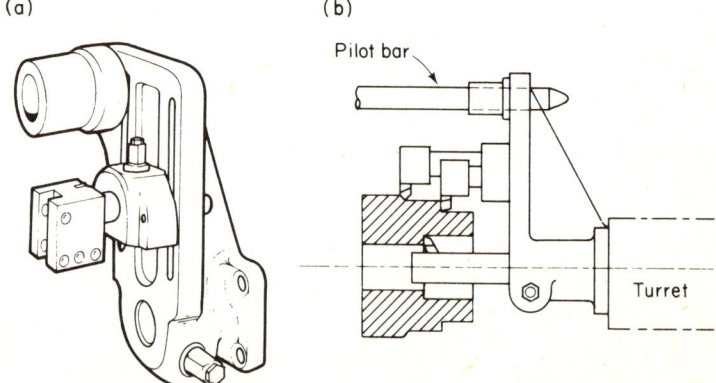

Figure 5.8 (a) Knee-turning toolholder, (b) use of combination knee tooling (H. W. Ward & Co)

Tooling for drilling, reaming and boring

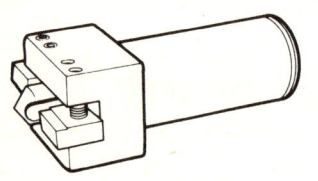

Figure 5.9 Start-drilling toolholder (H. W. Ward & Co)

Drilling in capstan and turret lathes is carried out in a similar fashion as for centre lathes except that the drill chuck is attached to and operated from the turret. Work is often prepared for drilling by using a *start-drilling toolholder* (Figure 5.9) which is also provided with a pair of cutters to enable operations such as facing and chamfering to be performed.

Reaming is carried out from the turret, the reamer generally being mounted in a special floating adaptor which permits the reamer to centralise itself in the hole. If the reamer is attached rigidly to the turret then an inaccurate hold could result if the turret alignment is not true with the work axis.

Boring is also performed from the turret using boring bars employing single or multiple tools (Figure 5.10(a)). Where long boring bars are used these are provided with a front pilot which engages with a pilot bush in the machine spindle to provide greater rigidity, as shown in Figure 5.10(b).

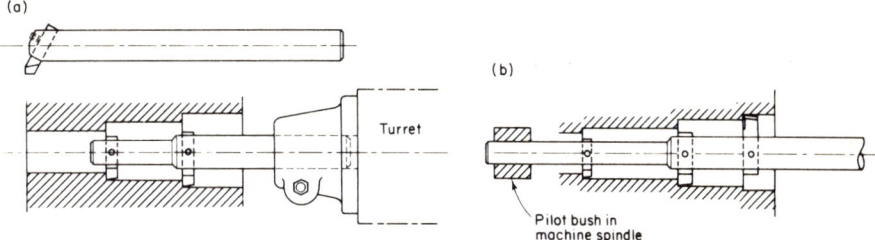

Figure 5.10 Boring operations

Threading The majority of threading operations are carried out using a diehead attached to the turret. The diehead contains four chaser-type cutting tools which are interchangeable to suit thread forms and pitches. In operation the diehead is presented to the work which, after making contact, will screw itself along the prepared diameter. On reaching the end of its travel the chasers will retract clear of the thread diameter to enable the diehead to be withdrawn rapidly. A typical diehead is shown in Figure 5.11.

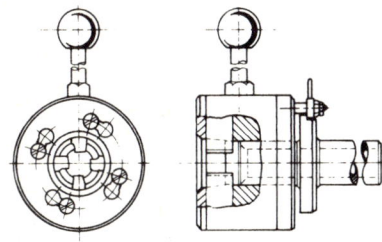

Figure 5.11 Diehead (H. W. Ward & Co)

Internal threads may be produced by using conventional or collapsible taps. The latter are preferred since they may be withdrawn without reversing the machine. They are confined, however, to thread diameters of about 20 mm or more. Large-diameter threads (external and internal) may be machined by chasing, the chaser movement being controlled by a headscrew in a fashion similar to screw cutting on a centre lathe.

Taper-turning Short *tapers* and *chamfers* may be formed from tools on the cross-slide. For longer tapers, external or internal, a taper-turning attachment is used (Figure 5.12). This consists of a horizontal slide, fitted to one of the turret faces, from which projects an arm fitted at its end with a roller which engages in a guide inclined to the axis of the machine. Therefore, as the

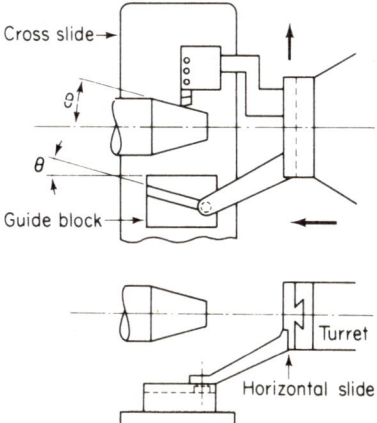

Figure 5.12 Taper-turning attachment

turret moves forward the tool attached to the horizontal slide will move parallel to the guide, thus producing a taper on the work.

Knurling Knurling is carried out from the turret using a tool illustrated in Figure 5.13. Knurling may be either straight or diagonal. For diagonal knurling the wheels are inclined at 45° to each other.

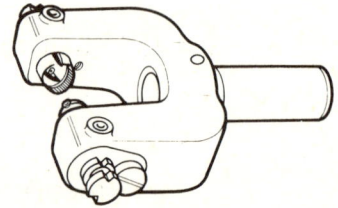

Figure 5.13 Knurling toolholder (H. W. Ward & Co)

PROCESS PLANNING The planning of work to be produced on capstan and turret lathes requires careful thought. For simple components the

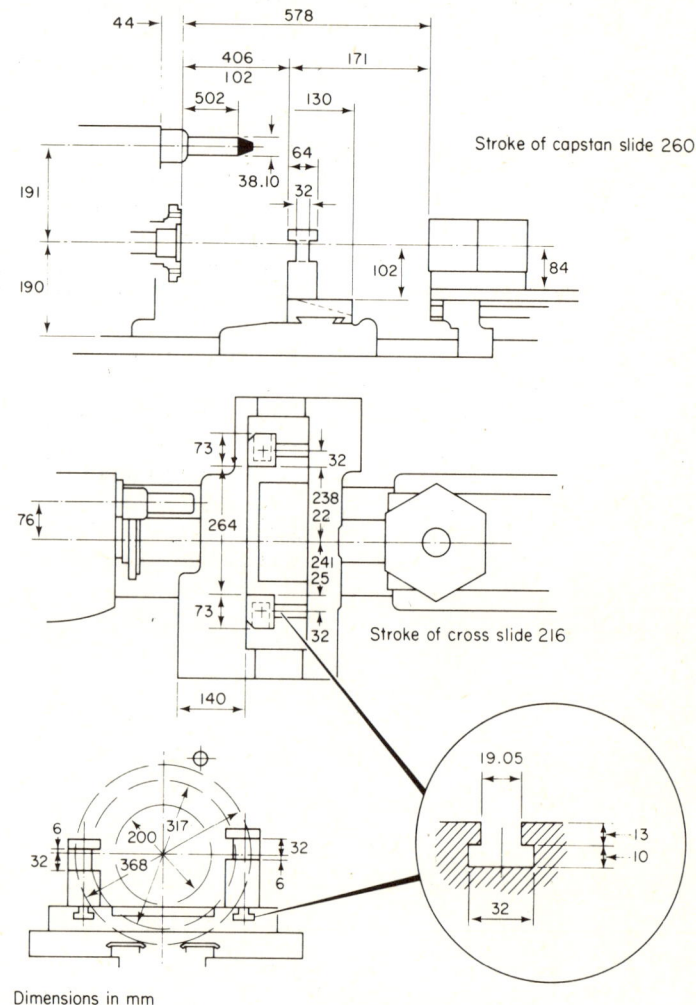

Dimensions in mm

Figure 5.14 Capstan lathe capacity chart (H. W. Ward & Co)

planning may be left to the toolsetter. However, for more involved workpieces planning is carried out by the planning engineer who will have a specialised knowledge regarding the use of capstan and turret lathes.

In planning work for these machines a tool layout is produced showing the types of tools required and the sequence in which they are to be used. When preparing these layouts it is usual to refer to the 'capacity chart' of the machine to ensure that tool movements, turret indexings, etc. clear the various machine parts. A typical capacity chart is illustrated in Figure 5.14. Furthermore, by listing the operations required systematically in the form of a tool layout it is possible to establish the machining cycle time and hence the cost per component.

Machining times The machining times for turning and drilling operations may be found from the following:

Turning time (min) = $\dfrac{L}{F \times N}$ where L = length to be cut (min)
N = spindle speed (rev/min)
F = feed (mm/rev)

Drilling time (min) = $\dfrac{D}{F \times N}$ where D = depth to be drilled (min)
F = feed (mm/rev)
N = spindle speed (rev/min)

Standard times During the machining of a component, the operator will have to make certain manipulations such as indexing the turret, feeding bar to stop, etc. These non-cutting operations have been assigned time allowances known as 'standard times'. An example of a range of standard times related to capstan and turret lathes is shown below.

Manipulation	Standard time (min)
Feed to bar stop	0.08
Index turret	0.08
Change speed	0.05
Change feed	0.05

Fatigue allowance Due to the repetitive nature of capstan and turret lathe work it is usual to add a fatigue allowance to operating times. This allowance varies slightly and depends on the machining conditions, but a typical allowance would be about 12% of the operating time.

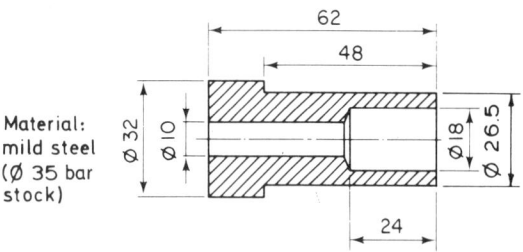

Material: mild steel (Ø 35 bar stock)

Tolerance on all dimensions ±0.10
Dimensions in mm

Figure 5.15 Shoulder spigot

102 Capstan and turret lathes

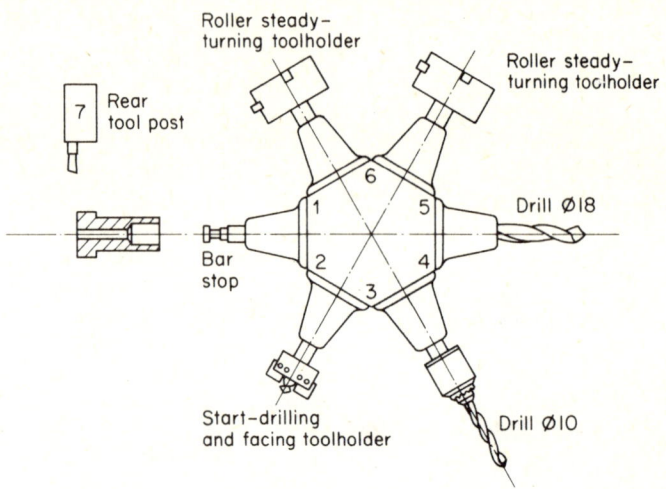

Operation number	Description of operation	Tooling station	Spindle speed (rev/min)	Feed (mm/rev)	Time (min)
1	Feed to bar stop	Turret 1	—	—	0.08
2	Start drill and face	Turret 2	710	—	0.20
3	Drill ⌀ 10	Turret 3	710	0.15	0.60
4	Drill ⌀ 18 × 24 deep	Turret 4	350	0.25	0.27
5	Turn ⌀ 32 × 63 long	Turret 5	254	0.30	0.83
6	Turn ⌀ 26.5 × 48 long	Turret 6	254	0.30	0.63
7	Part off	Rear tool post	254	0.15	0.33

Cutting time   2.94 min
Manipulation time   0.68 min
Allowance (12%)   0.43 min
Overall machining time   4.0 min

Figure 5.16 Tool layout

Example: Prepare a tool layout for the component shown in Figure 5.15. The format of the layout is shown in Figure 5.16, the spindle speeds and feeds having been obtained from references.

Operation 1   *Feed to bar stop*
                 Standard time = **0.08 min**

Operation 2   *Start drill and face*
                 This time is difficult to calculate precisely and is therefore given an estimated time of **0.20 min**

Operation 3   *Drill ⌀ 10 mm*
                 To ensure that the component parts off, the depth of the hole is taken to 63 mm

$$\text{Time} = \frac{D}{F \times N} = \frac{63}{0.15 \times 710} = \mathbf{0.60\ min}$$

Operation 4   *Drill ⌀ 18 mm × 24 mm long*

$$\text{Time} = \frac{D}{F \times N} = \frac{24}{0.25 \times 350} = \mathbf{0.27\ min}$$

> **Operation 5**   *Turn to ⌀ 32 mm × 63 mm long*
> To ensure that the component parts off, the length is turned to 63 mm
>
> $$\text{Time} = \frac{L}{F \times N} = \frac{63}{0.3 \times 254} = \mathbf{0.83\ min}$$
>
> **Operation 6**   *Turn to ⌀ 26.5 mm × 48 mm long*
>
> $$\text{Time} = \frac{L}{F \times N} = \frac{48}{0.3 \times 254} = \mathbf{0.63\ min}$$
>
> **Operation 7**   *Part off*
> For this operation the tool is required to feed in 27 mm. Parting-off operations are carried out using a slower feed.
>
> $$\text{Time} = \frac{L}{F \times N} = \frac{12.5}{0.15 \times 254} = \mathbf{0.33\ min}$$
>
> $$\text{Cutting time} = 0.08 + 0.20 + 0.60 + 0.27 + 0.83 + 0.63 + 0.33 = \mathbf{2.94\ min}$$
>
> *Manipulation times*
>
> (1) There are six turret indexings
>   therefore, time = 6 × 0.08 = **0.48 min**
>
> (2) There are two feed changes
>   therefore, time = 2 × 0.05 = **0.10 min**
>
> (3) There are two speed changes
>   therefore, time = 2 × 0.05 = **0.10 min**
>
> Therefore, total manipulation time = 0.48 + 0.10 + 0.10
>   = **0.68 min**
>
> Machining time = cutting time + manipulation time
>   = 2.94 + 0.68
>   = 3.62 min
>
> Therefore, overall machining time, including fatigue allowance,
>
> becomes $\quad 3.62 + \left(3.62 \times \dfrac{12}{100}\right)$
>   = 3.62 + 0.43
>   = **4.0 min**

MACHINING COSTS

The total cost of machining a component is controlled by the following three factors:

(a) Operator's wage, based on an hourly rate.
(b) Time taken to machine the component.
(c) Factory overheads.

The factory overheads are the costs not directly attributable to the machining of the component, e.g. lighting, heating, salaries of office staff, etc. It is usual to allocate a fixed value for overheads, this value being very much dependent on the manufacturing organisation concerned. However, in all cases the cost of overheads will be found to be fairly high and is usually expressed as a percentage of the operator cost.

Therefore, total machining cost = operator cost + overhead costs.

> *Example:* A capstan operator earning £1.80 per hour produces components at a rate of one every 4.5 minutes. Assuming the factory overheads are 350%, determine the total machining cost for this component.
>
> $$\text{Operator cost per component} = £1.80 \times \frac{4.5}{60}$$
> $$= £0.135$$
> $$\text{Overhead cost per component} = £0.135 \times \frac{350}{100}$$
> $$= £0.47$$
> $$\text{Total machining cost} = 0.135 + 0.47$$
> $$= \mathbf{£0.60}$$

SUMMARY

Capstan and turret lathes are used for producing parts rapidly using pre-set toolholders. Capstan lathes are used mainly for bar work while turret lathes are generally used for larger work often in the form of castings and forgings. The tool travel on these machines is controlled by a series of stops. Work-holding arrangements include collet and jaw chucks (two- or three-jaw) which may be used with soft jaws to accommodate irregularly-shaped work. These chucks may be operated mechanically, hydraulically or electrically. To ensure that capstan and turret lathes are used to the best economic advantage tool layouts are prepared from which the overall machining time may be determined.

QUESTIONS

(1) Explain clearly the difference between a capstan lathe and a turret lathe.

(2) Explain with the aid of sketches the following types of toolholders:
    (i)   Roller steady-turning toolholder
    (ii) Knee-turning toolholder
    (iii) Start-turning toolholder.

(3) Describe how external and internal threads are normally produced on capstan and turret lathes.

(4) Using a suitable sketch explain the function of 'soft jaws' for holding work in jaw chucks.

(5) With reference to process planning what is meant by standard times?

(6) Show by means of a sketch how a taper may be machined on a capstan or turret lathe.

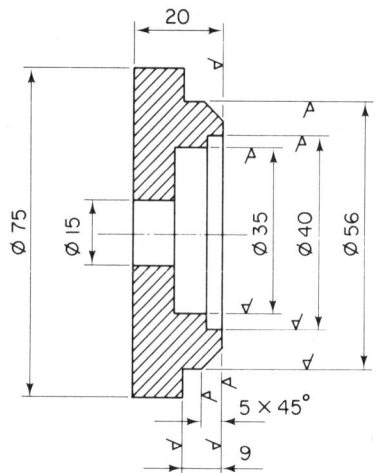

Machine where marked thus ✓
Tolerance on all dimensions ± 0.10
Dimensions in mm
Material: aluminium alloy

Figure 5.17 Bearing housing

(7) The machining cycle time for a component produced on a capstan lathe was found to be 5.3 minutes. If the operator's hourly rate is £1.60 and the overheads are 350% determine the machining cost per component.

(8) The component shown in Figure 5.17 is to be produced on a capstan lathe. Prepare a tool layout and hence determine the overall machining time.